The Science Delusion

The Science Delusion

Freeing the Spirit of Enquiry

RUPERT SHELDRAKE

CORONET

First published in Great Britain in 2012 by Coronet
An imprint of Hodder & Stoughton
An Hachette UK company

3

Copyright © Rupert Sheldrake 2012

The right of Rupert Sheldrake to be identified as the Author of the Work has been
asserted by him in accordance with the Copyright, Designs and Patents Act 1988.

A CIP catalogue record for this title is available from the British Library

Hardback ISBN 978 1 444 72792 0
Trade Paperback ISBN 978 1 444 72793 7
eBook ISBN 978 1 444 72795 1

Typeset in Minion by Hewer Text UK Ltd, Edinburgh

Printed and bound by CPI Group (UK) Ltd, Croydon, CR0 4YY

Hodder & Stoughton policy is to use papers that are natural, renewable
and recyclable products and made from wood grown in sustainable
forests. The logging and manufacturing processes are expected to
conform to the environmental regulations of the country of origin.

Hodder & Stoughton Ltd
338 Euston Road
London NW1 3BH

www.hodder.co.uk

For all those who have helped and encouraged me, especially my wife Jill and our sons Merlin and Cosmo.

Contents

Preface

My interest in science began when I was very young. As a child I kept many kinds of animals, ranging from caterpillars and tadpoles to pigeons, rabbits, tortoises and a dog. My father, a herbalist, pharmacist and microscopist, taught me about plants from my earliest years. He showed me a world of wonders through his microscope, including tiny creatures in drops of pond water, scales on butterflies' wings, shells of diatoms, cross-sections of plant stems and a sample of radium that glowed in the dark. I collected plants and read books on natural history, like Fabre's *Book of Insects*, which told the life stories of scarab beetles, praying mantises and glow-worms. By the time I was twelve years old I wanted to become a biologist.

I studied sciences at school and then at Cambridge University, where I majored in biochemistry. I liked what I was doing, but found the focus very narrow, and wanted to see a bigger picture. I had a life-changing opportunity to widen my perspective when I was awarded a Frank Knox fellowship in the graduate school at Harvard, where I studied the history and philosophy of science.

I returned to Cambridge to do research on the development of plants. In the course of my PhD project, I made an original discovery: dying cells play a major part in the regulation of plant growth, releasing the plant hormone auxin as they break down in the process of 'programmed cell death'. Inside growing plants, new wood cells dissolve themselves as they die, leaving their cellulose walls as microscopic tubes through which water is conducted in stems, roots and veins of leaves. I discovered that auxin is produced as cells die,[1] that dying cells stimulate more growth; more growth leads to more death, and hence to more growth.

After receiving my PhD, I was elected to a research fellowship of Clare College, Cambridge, where I was director of studies in cell biology and biochemistry, teaching students in tutorials and lab classes. I was then appointed a research fellow of the Royal Society and continued my research at Cambridge on plant hormones, studying the way in which auxin is transported from the shoots towards the root tips. With my colleague Philip Rubery, I worked out the molecular basis of polar auxin transport,[2] providing a foundation on which much subsequent research on plant polarity has been built.

Funded by the Royal Society, I spent a year at the University of Malaya, studying rain forest ferns, and at the Rubber Research Institute of Malaya I discovered how the flow of latex in rubber trees is regulated genetically, and I shed new light on the development of latex vessels.[3]

When I returned to Cambridge, I developed a new hypothesis of ageing in plants and animals, including humans. All cells age. When they stop growing, they eventually die. My hypothesis is about rejuvenation, and proposes that harmful waste products build up in all cells, causing them to age, but they can produce rejuvenated daughter cells by asymmetric cell divisions in which one cell receives most of these waste products and is doomed, while the other is wiped clean. The most rejuvenated of all cells are eggs. In both plants and animals, two successive cell divisions (meiosis) produce an egg cell and three sister cells, which quickly die. My hypothesis was published in *Nature* in 1974 in a paper called 'The ageing, growth and death of cells'.[4] 'Programmed cell death', or 'apoptosis', has since become a major field of research, important for our understanding of diseases such as cancer and HIV, as well as tissue regeneration through stem cells. Many stem cells divide asymmetrically, producing a new, rejuvenated stem cell and a cell that differentiates, ages and dies. My hypothesis is that the rejuvenation of stem cells through cell division depends on their sisters paying the price of mortality.

Wanting to broaden my horizons and do practical research that could benefit some of the world's poorest people, I left Cambridge to join the International Crops Research Institute for the Semi-Arid Tropics, near Hyderabad, India, as Principal Plant Physiologist, working on chickpeas and pigeonpeas.[5] We bred new high-yielding varieties of these crops, and developed multiple cropping systems[6] that are now widely used by farmers in Asia and Africa, greatly increasing yields.

A new phase in my scientific career began in 1981 with the publication of my book *A New Science of Life*, in which I suggested a hypothesis of form-shaping fields, called morphogenetic fields, that control the development of animal embryos and the growth of plants. I proposed that these fields have an inherent memory, given by a process called morphic resonance. This hypothesis was supported by the available evidence and gave rise to a range of experimental tests, summarised in the new edition of *A New Science of Life* (2009).

After my return to England from India, I continued to investigate plant development, and also started research with homing pigeons, which had intrigued me since I kept pigeons as a child. How do pigeons find their way home from hundreds of miles away, across unfamiliar terrain and even across the sea? I thought they might be linked to their home by a field that acted like an invisible elastic band, pulling them homewards. Even if they have a magnetic sense as well, they cannot find their home just by knowing compass directions. If you were parachuted into unknown territory with a compass, you would know where north was, but not where your home was.

I came to realise that pigeon navigation was just one of many unexplained powers of animals. Another was the ability of some dogs to know when their owners are coming home, seemingly telepathically. It was not difficult or expensive to do research on these subjects, and the results were fascinating. In 1994 I published a book called *Seven Experiments that Could Change the World* in which I proposed low-cost tests that could change our ideas about the nature of reality, with results that were summarised in a new edition (2002), and in my books *Dogs That Know When Their*

Owners Are Coming Home (1999; new edition 2011) and *The Sense of Being Stared At* (2003).

For the last twenty years I have been a Fellow of the Institute of Noetic Sciences, near San Francisco, and a visiting professor at several universities, including the Graduate Institute in Connecticut. I have published more than eighty papers in peer-reviewed scientific journals, including several in *Nature*. I belong to a range of scientific societies, including the Society for Experimental Biology and the Society for Scientific Exploration, and I am a fellow of the Zoological Society and the Cambridge Philosophical Society. I give seminars and lectures on my research at a wide variety of universities, research institutes and scientific conferences in Britain, continental Europe, North and South America, India and Australasia.

I have spent all my adult life as a scientist, and I strongly believe in the importance of the scientific approach. Yet I have become increasingly convinced that the sciences have lost much of their vigour, vitality and curiosity. Dogmatic ideology, fear-based conformity and institutional inertia are inhibiting scientific creativity.

With scientific colleagues, I have been struck over and over again by the contrast between public and private discussions. In public, scientists are very aware of the powerful taboos that restrict the range of permissible topics; in private they are often more adventurous.

I have written this book because I believe that the sciences will be more exciting and engaging when they move beyond the dogmas that restrict free enquiry and imprison imaginations.

Many people have contributed to these explorations through discussions, debates, arguments and advice, and I cannot begin to mention everyone to whom I am indebted. This book is dedicated to all those who have helped and encouraged me.

I am grateful for the financial support that has enabled me to write this book: from Trinity College, Cambridge, where I was the Perrott-Warrick Senior Researcher from 2005 to 2010; from Addison Fischer and the Planet Heritage Foundation; and from the Watson Family Foundation and the Institute of Noetic

Sciences. I also thank my research assistant, Pamela Smart, and my webmaster, John Caton, for their much-appreciated help.

This book has benefited from many comments on drafts. In particular, I thank Bernard Carr, Angelika Cawdor, Nadia Cheney, John Cobb, Ted Dace, Larry Dossey, Lindy Dufferin and Ava, Douglas Hedley, Francis Huxley, Robert Jackson, Jürgen Krönig, James Le Fanu, Peter Fry, Charlie Murphy, Jill Purce, Anthony Ramsay, Edward St Aubyn, Cosmo Sheldrake, Merlin Sheldrake, Jim Slater, Pamela Smart, Peggy Taylor and Christoffer van Tulleken as well as my agent Jim Levine, in New York, and my editor at Hodder and Stoughton, Mark Booth.

Introduction

THE TEN DOGMAS OF MODERN SCIENCE

The 'scientific worldview' is immensely influential because the sciences have been so successful. They touch all our lives through technologies and through modern medicine. Our intellectual world has been transformed by an immense expansion of knowledge, down into the most microscopic particles of matter and out into the vastness of space, with hundreds of billions of galaxies in an ever-expanding universe.

Yet in the second decade of the twenty-first century, when science and technology seem to be at the peak of their power, when their influence has spread all over the world and when their triumph seems indisputable, unexpected problems are disrupting the sciences from within. Most scientists take it for granted that these problems will eventually be solved by more research along established lines, but some, including myself, think they are symptoms of a deeper malaise.

In this book, I argue that science is being held back by centuries-old assumptions that have hardened into dogmas. The sciences would be better off without them: freer, more interesting, and more fun.

The biggest scientific delusion of all is that science already knows the answers. The details still need working out but, in principle, the fundamental questions are settled.

Contemporary science is based on the claim that all reality is material or physical. There is no reality but material reality. Consciousness is a by-product of the physical activity of the brain. Matter is unconscious. Evolution is purposeless. God exists only as an idea in human minds, and hence in human heads.

These beliefs are powerful, not because most scientists think about them critically but because they don't. The *facts* of science are real enough; so are the techniques that scientists use, and the technologies based on them. But the belief system that governs conventional scientific thinking is an act of faith, grounded in a nineteenth-century ideology.

This book is pro-science. I want the sciences to be less dogmatic and more scientific. I believe that the sciences will be regenerated when they are liberated from the dogmas that constrict them.

The scientific creed

Here are the ten core beliefs that most scientists take for granted.

1. Everything is essentially mechanical. Dogs, for example, are complex mechanisms, rather than living organisms with goals of their own. Even people are machines, 'lumbering robots', in Richard Dawkins's vivid phrase, with brains that are like genetically programmed computers.
2. All matter is unconscious. It has no inner life or subjectivity or point of view. Even human consciousness is an illusion produced by the material activities of brains.
3. The total amount of matter and energy is always the same (with the exception of the Big Bang, when all the matter and energy of the universe suddenly appeared).
4. The laws of nature are fixed. They are the same today as they were at the beginning, and they will stay the same for ever.
5. Nature is purposeless, and evolution has no goal or direction.
6. All biological inheritance is material, carried in the genetic material, DNA, and in other material structures.
7. Minds are inside heads and are nothing but the activities of brains. When you look at a tree, the image of the tree you are seeing is not 'out there', where it seems to be, but inside your brain.
8. Memories are stored as material traces in brains and are wiped out at death.

9. Unexplained phenomena like telepathy are illusory.
10. Mechanistic medicine is the only kind that really works.

Together, these beliefs make up the philosophy or ideology of materialism, whose central assumption is that everything is essentially material or physical, even minds. This belief-system became dominant within science in the late nineteenth century, and is now taken for granted. Many scientists are unaware that materialism is an assumption: they simply think of it as science, or the scientific view of reality, or the scientific worldview. They are not actually taught about it, or given a chance to discuss it. They absorb it by a kind of intellectual osmosis.

In everyday usage, materialism refers to a way of life devoted entirely to material interests, a preoccupation with wealth, possessions and luxury. These attitudes are no doubt encouraged by the materialist philosophy, which denies the existence of any spiritual realities or non-material goals, but in this book I am concerned with materialism's scientific claims, rather than its effects on lifestyles.

In the spirit of radical scepticism, I turn each of these ten doctrines into a question. Entirely new vistas open up when a widely accepted assumption is taken as the beginning of an enquiry, rather than as an unquestionable truth. For example, the assumption that nature is machine-like or mechanical becomes a question: 'Is nature mechanical?' The assumption that matter is unconscious becomes 'Is matter unconscious?' And so on.

In the Prologue I look at the interactions of science, religion and power, and then in Chapters 1 to 10, I examine each of the ten dogmas. At the end of each chapter, I discuss what difference this topic makes and how it affects the way we live our lives. I also pose several further questions, so that any readers who want to discuss these subjects with friends or colleagues will have some useful starting points. Each chapter is followed by a summary.

The credibility crunch for the 'scientific worldview'

For more than two hundred years, materialists have promised that science will eventually explain everything in terms of physics and chemistry. Science will prove that living organisms are complex machines, minds are nothing but brain activity and nature is purposeless. Believers are sustained by the faith that scientific discoveries will justify their beliefs. The philosopher of science Karl Popper called this stance 'promissory materialism' because it depends on issuing promissory notes for discoveries not yet made.[1] Despite all the achievements of science and technology, materialism is now facing a credibility crunch that was unimaginable in the twentieth century.

In 1963, when I was studying biochemistry at Cambridge University, I was invited to a series of private meetings with Francis Crick and Sydney Brenner in Brenner's rooms in King's College, along with a few of my classmates. Crick and Brenner had recently helped to 'crack' the genetic code. Both were ardent materialists and Crick was also a militant atheist. They explained there were two major unsolved problems in biology: development and consciousness. They had not been solved because the people who worked on them were not molecular biologists – or very bright. Crick and Brenner were going to find the answers within ten years, or maybe twenty. Brenner would take developmental biology, and Crick consciousness. They invited us to join them.

Both tried their best. Brenner was awarded the Nobel Prize in 2002 for his work on the development of a tiny worm, *Caenorhabdytis elegans*. Crick corrected the manuscript of his final paper on the brain the day before he died in 2004. At his funeral, his son Michael said that what made him tick was not the desire to be famous, wealthy or popular, but 'to knock the final nail into the coffin of vitalism'. (Vitalism is the theory that living organisms are truly alive, and not explicable in terms of physics and chemistry alone.)

Crick and Brenner failed. The problems of development and consciousness remain unsolved. Many details have been discovered, dozens of genomes have been sequenced, and brain scans are ever more precise. But there is still no proof that life and minds can be explained by physics and chemistry alone (see Chapters 1, 4 and 8).

The fundamental proposition of materialism is that matter is the only reality. Therefore consciousness is nothing but brain activity. It is either like a shadow, an 'epiphenomenon', that does nothing, or it is just another way of *talking* about brain activity. However, among contemporary researchers in neuroscience and consciousness studies there is no consensus about the nature of minds. Leading journals such as *Behavioural and Brain Sciences* and the *Journal of Consciousness Studies* publish many articles that reveal deep problems with the materialist doctrine. The philosopher David Chalmers has called the very existence of subjective experience the 'hard problem'. It is hard because it defies explanation in terms of mechanisms. Even if we understand how eyes and brains respond to red light, the *experience* of redness is not accounted for.

In biology and psychology the credibility rating of materialism is falling. Can physics ride to the rescue? Some materialists prefer to call themselves physicalists, to emphasise that their hopes depend on modern physics, not nineteenth-century theories of matter. But physicalism's own credibility rating has been reduced by physics itself, for four reasons.

First, some physicists insist that quantum mechanics cannot be formulated without taking into account the minds of observers. They argue that minds cannot be reduced to physics because physics presupposes the minds of physicists.[2]

Second, the most ambitious unified theories of physical reality, string and M-theories, with ten and eleven dimensions respectively, take science into completely new territory. Strangely, as Stephen Hawking tells us in his book *The Grand Design* (2010), 'No one seems to know what the "M" stands for, but it may be "master", "miracle" or "mystery"'. According to what Hawking calls 'model-dependent realism', different theories may have to be applied in different situations. 'Each theory may have its own version of reality, but according to model-dependent realism, that is acceptable so long as the theories agree in their predictions whenever they overlap, that is, whenever they can both be applied.'[3]

String theories and M-theories are currently untestable so 'model-dependent realism' can only be judged by reference to other models, rather than by experiment. It also applies to countless other universes, none of which has ever been observed. As Hawking points out,

> M-theory has solutions that allow for *different universes* with different apparent laws, depending on how the internal space is curled. M-theory has solutions that allow for many different internal spaces, perhaps as many as 10^{500}, which means it allows for 10^{500} different universes, each with its own laws . . . The original hope of physics to produce a single theory explaining the apparent laws of our universe as the unique possible consequence of a few simple assumptions may have to be abandoned.[4]

Some physicists are deeply sceptical about this entire approach, as the theoretical physicist Lee Smolin shows in his book *The Trouble With Physics: The Rise of String Theory, the Fall of a Science and What Comes Next* (2008).[5] String theories, M-theories and 'model-dependent realism' are a shaky foundation for materialism or physicalism or any other belief system, as discussed in Chapter 1.

Third, since the beginning of the twenty-first century, it has become apparent that the known kinds of matter and energy make up only about four per cent of the universe. The rest consists of 'dark matter' and 'dark energy'. The nature of 96 per cent of physical reality is literally obscure (see Chapter 2).

Fourth, the Cosmological Anthropic Principle asserts that if the laws and constants of nature had been slightly different at the moment of the Big Bang, biological life could never have emerged, and hence we would not be here to think about it (see Chapter 3). So did a divine mind fine-tune the laws and constants in the beginning? To avoid a creator God emerging in a new guise, most leading cosmologists prefer to believe that our universe is one of a vast, and perhaps infinite, number of parallel universes, all with different laws and constants, as M-theory also suggests. We just happen to exist in the one that has the right conditions for us.[6]

This multiverse theory is the ultimate violation of Occam's Razor, the philosophical principle that 'entities must not be multiplied beyond necessity', or in other words, that we should make as few assumptions as possible. It also has the major disadvantage of being untestable.[7] And it does not even succeed in getting rid of God. An infinite God could be the God of an infinite number of universes.[8]

Materialism provided a seemingly simple, straightforward worldview in the late nineteenth century, but twenty-first-century science has left it behind. Its promises have not been fulfilled, and its promissory notes have been devalued by hyperinflation.

I am convinced that the sciences are being held back by assumptions that have hardened into dogmas, maintained by powerful taboos. These beliefs protect the citadel of established science, but act as barriers against open-minded thinking.

Prologue

SCIENCE, RELIGION AND POWER

Since the late nineteenth century, science has dominated and transformed the earth. It has touched everyone's lives through technology and modern medicine. Its intellectual prestige is almost unchallenged. Its influence is greater than that of any other system of thought in all of human history. Although most of its power comes from its practical applications, it also has a strong intellectual appeal. It offers new ways of understanding the world, including the mathematical order at the heart of atoms and molecules, the molecular biology of genes, and the vast sweep of cosmic evolution.

The scientific priesthood

Francis Bacon (1561–1626), a politician and lawyer who became Lord Chancellor of England, foresaw the power of organised science more than anyone else. To clear the way, he needed to show that there was nothing sinister about acquiring power over nature. When he was writing, there was a widespread fear of witchcraft and black magic, which he tried to counteract by claiming that knowledge of nature was God-given, not inspired by the devil. Science was a return to the innocence of the first man, Adam, in the Garden of Eden before the Fall.

Bacon argued that the first book of the Bible, Genesis, justified scientific knowledge. He equated man's knowledge of nature with Adam's naming of the animals. God 'brought them unto Adam to see what he would call them, and what Adam called every living creature, that was the name thereof'. (Genesis 2: 19–20) This was literally man's knowledge, because Eve was not created until two verses later. Bacon argued that man's technological mastery of nature was the recovery of a God-given power, rather than something new. He confidently assumed that people would use their new knowledge wisely and well: 'Only let the human race recover that right over nature which belongs to it by divine bequest; the exercise thereof will be governed by sound reason and true religion.'[1]

The key to this new power over nature was organised institutional research. In *New Atlantis* (1624), Bacon described a technocratic Utopia in which a scientific priesthood made decisions for the good of the state as a whole. The Fellows of this scientific 'Order or Society' wore long robes and were treated with a respect that their power and dignity required. The head of the order travelled in a rich chariot, under a radiant golden image of the sun. As he rode in procession, 'he held up his bare hand, as he went, as blessing the people'.

The general purpose of this foundation was 'the knowledge of causes and secret motions of things; and the enlarging of human empire, to the effecting of all things possible'. The Society was equipped with machinery and facilities for testing explosives and armaments, experimental furnaces, gardens for plant breeding, and dispensaries.[2]

This visionary scientific institution foreshadowed many features of institutional research, and was a direct inspiration for the founding of the Royal Society in London in 1660, and for many other national academies of science. But although the members of these academies were often held in high esteem, none achieved the grandeur and political power of Bacon's imaginary prototypes. Their glory was continued even after their deaths in a gallery, like a Hall of Fame, where their images were preserved. 'For upon every invention of value we erect a statue to the inventor, and give him a liberal and honourable reward.'[3]

In England in Bacon's time (and still today) the Church of England was linked to the state as the Established Church. Bacon envisaged that the scientific priesthood would also be linked to the state through state patronage, forming a kind of established church of science. And here again he was prophetic. In nations both capitalist and Communist, the official academies of science remain the centres of power of the scientific establishment. There is no separation of science and state. Scientists play the role of an established priesthood, influencing government policies on the arts of warfare, industry, agriculture, medicine, education and research.

Bacon coined the ideal slogan for soliciting financial support from governments and investors: 'Knowledge is power.'[4] But the success of scientists in eliciting funding from governments varied from country to country. The systematic state funding of science began much earlier in France and Germany than in Britain and the United States where, until the latter half of the nineteenth century, most research was privately funded or carried out by wealthy amateurs like Charles Darwin.[5]

In France, Louis Pasteur (1822–95) was an influential proponent of science as a truth-finding religion, with laboratories like temples through which mankind would be elevated to its highest potential:

> Take interest, I beseech you, in those sacred institutions which we designate under the expressive name of laboratories. Demand that they be multiplied and adorned; they are the temples of wealth and of the future. There it is that humanity grows, becomes stronger and better.[6]

By the beginning of the twentieth century, science was almost entirely institutionalised and professionalised, and after the Second World War expanded enormously under government patronage, as well as through corporate investment.[7] The highest level of funding is in the United States, where in 2008 the total expenditure on research and development was $398 billion, of which $104 billion came from the government.[8] But governments and corporations do not usually pay scientists to do research because they want innocent knowledge, like that of Adam before the Fall. Naming animals, as in classifying endangered species of beetles in tropical rainforests, is a low priority.

Most funding is a response to Bacon's persuasive slogan 'knowledge is power'.

By the 1950s, when institutional science had reached an unprecedented level of power and prestige, the historian of science George Sarton approvingly described the situation in a way that sounds like the Roman Catholic Church before the Reformation:

> Truth can be determined only by the judgement of experts ...
> Everything is decided by very small groups of men, in fact, by
> single experts whose results are carefully checked, however, by a
> few others. The people have nothing to say but simply to accept the
> decisions handed out to them. Scientific activities are controlled by
> universities, academies and scientific societies, but such control is
> as far removed from popular control as it possibly could be.[9]

Bacon's vision of a scientific priesthood has now been realised on a global scale. But his confidence that man's power over nature would be guided by 'sound reason and true religion' was misplaced.

The fantasy of omniscience

The fantasy of omniscience is a recurrent theme in the history of science, as scientists aspire to a total godlike knowledge. At the beginning of the nineteenth century, the French physicist Pierre-Simon Laplace imagined a scientific mind capable of knowing and predicting everything:

> Consider an intelligence which, at any instant, could have a
> knowledge of all the forces controlling nature together with the
> momentary conditions of all the entities of which nature consists.
> If this intelligence were powerful enough to submit all these data
> to analysis it would be able to embrace in a single formula the
> movements of the largest bodies in the universe and those of the
> lightest atoms; for it nothing would be uncertain; the past and
> future would be equally present for its eyes.[10]

These ideas were not confined to physicists. Thomas Henry Huxley, who did so much to propagate Darwin's theory of evolution, extended mechanical determinism to cover the entire evolutionary process:

> If the fundamental proposition of evolution is true, that the entire world, living and not living, is the result of the mutual interaction, according to definite laws, of the forces possessed by the molecules of which the primitive nebulosity of the universe was composed, it is no less certain the existing world lay, potentially, in the cosmic vapour, and that a sufficient intellect could, from a knowledge of the properties of the molecules of that vapour, have predicted, say, the state of the fauna of Great Britain in 1869.[11]

When the belief in determinism was applied to the activity of the human brain, it resulted in a denial of free will, on the grounds that everything about the molecular and physical activities of the brain was in principle predictable. Yet this conviction rested not on scientific evidence, but simply on the *assumption* that everything was fully determined by mathematical laws.

Even today, many scientists assume that free will is an illusion. Not only is the activity of the brain determined by machine-like processes, but there is no non-mechanical self capable of making choices. For example, in 2010, the British brain scientist Patrick Haggard asserted, 'As a neuroscientist, you've got to be a determinist. There are physical laws, which the electrical and chemical events in the brain obey. Under identical circumstances, you couldn't have done otherwise. There's no "I" which can say, "I want to do otherwise." '[12] However, Haggard does not let his scientific beliefs interfere with his personal life: 'I keep my scientific and personal lives pretty separate. I still seem to decide what films I go to see, I don't feel it's predestined, though it must be determined somewhere in my brain.'

Indeterminism and chance

In 1927, with the recognition of the uncertainty principle in quantum physics, it became clear that indeterminism was an essential feature of the physical world, and physical predictions could be made only in terms of probabilities. The fundamental reason is that quantum phenomena are wavelike, and a wave is by its very nature spread out in space and time: it cannot be localised at a single point at a particular instant; or, more technically, its position and momentum cannot both be known precisely.[13] Quantum theory deals in statistical probabilities, not certainties. The fact that one possibility is realised in a quantum event rather than another is a matter of chance.

Does quantum indeterminism affect the question of free will? Not if indeterminism is purely random. Choices made at random are no freer than if they are fully determined.[14]

In neo-Darwinian evolutionary theory randomness plays a central role through the chance mutations of genes, which are quantum events. With different chance events, evolution would happen differently. T. H. Huxley was wrong in believing that the course of evolution was predictable. 'Replay the tape of life,' said the evolutionary biologist Stephen Jay Gould, 'and a different set of survivors would grace our planet today.'[15]

In the twentieth century it became clear that not just quantum processes but almost all natural phenomena are probabilistic, including the turbulent flow of liquids, the breaking of waves on the seashore, and the weather: they show a spontaneity and indeterminism that eludes exact prediction. Weather forecasters still get it wrong in spite of having powerful computers and a continuous stream of data from satellites. This is not because they are bad scientists but because weather is intrinsically unpredictable in detail. It is chaotic, not in the everyday sense that there is no order at all, but in the sense that it is not precisely predictable. To some extent, the weather can be modelled mathematically in terms of chaotic dynamics, sometimes called 'chaos theory', but these models do not make exact predictions.[16] Certainty is as unachievable in the everyday world as it is in quantum physics. Even the orbits of the planets around the sun, long considered the centrepiece of mechanistic science, turn out to be chaotic over long time scales.[17]

The belief in determinism, strongly held by many nineteenth- and early-twentieth-century scientists, turned out to be a delusion. The freeing of scientists from this dogma led to a new appreciation of the indeterminism of nature in general, and of evolution in particular. The sciences have not come to an end by abandoning the belief in determinism. Likewise, they will survive the loss of the dogmas that still bind them; they will be regenerated by new possibilities.

Further fantasies of omniscience

By the end of the nineteenth century, the fantasy of scientific omniscience went far beyond a belief in determinism. In 1888, the Canadian-American astronomer Simon Newcomb wrote, 'We are probably nearing the limit of all we can know about astronomy.' In 1894, Albert Michelson, later to win the Nobel Prize for Physics, declared, 'The more important fundamental laws and facts of physical science have all been discovered, and these are now so firmly established that the possibility of their ever being supplanted in consequence of new discoveries is exceedingly remote . . . Our future discoveries must be looked for in the sixth place of decimals.'[18] And in 1900 William Thomson, Lord Kelvin, the physicist and inventor of intercontinental telegraphy, expressed this supreme confidence in an often-quoted (although perhaps apocryphal) claim: 'There is nothing new to be discovered in physics now. All that remains is more and more precise measurement.'

These convictions were shattered in the twentieth century through quantum physics, relativity theory, nuclear fission and fusion (as in atom and hydrogen bombs), the discovery of galaxies beyond our own, and the Big Bang theory – the idea that the universe began very small and very hot some 14 billion years ago and has been growing, cooling and evolving ever since.

Nevertheless, by the end of the twentieth century, the fantasy of omniscience was back again, this time fuelled by the triumphs of twentieth-century physics and by the discoveries of neurobiology and molecular biology. In 1997, John Horgan, a senior science writer at *Scientific American*, published a book called *The End of Science: Facing the Limits of Knowledge in the Twilight of the Scientific Age*. After interviewing many leading scientists, he advanced a provocative thesis:

> If one believes in science, one must accept the possibility – even the probability – that the great era of scientific discovery is over. By science I mean not applied science, but science at its purest and greatest, the primordial human quest to understand the universe and our place in it. Further research may yield no more great revelations or revolutions, but only incremental, diminishing returns.[19]

Horgan is surely right that once something has been discovered – like the structure of DNA – it cannot go on being discovered. But he took it for granted that the tenets of conventional science are true. He assumed that the most fundamental answers are already known. They are not, and every one of them can be replaced by more interesting and fruitful questions, as I show in this book.

Science and Christianity

The founders of mechanistic science in the seventeenth century, including Johannes Kepler, Galileo Galilei, René Descartes, Francis Bacon, Robert Boyle and Isaac Newton, were all practising Christians. Kepler, Galileo and Descartes were Roman Catholics; Bacon, Boyle and Newton Protestants. Boyle, a wealthy aristocrat, was exceptionally devout, and spent large amounts of his own money to promote missionary activity in India. Newton devoted much time and energy to biblical scholarship, with a particular interest in the dating of prophecies. He calculated that the Day of Judgment would occur between the years 2060 and 2344, and set out the details in his book *Observations on the Prophecies of Daniel and the Apocalypse of St John*.[20]

Seventeenth-century science created a vision of the universe as a machine intelligently designed and started off by God. Everything was governed by eternal mathematical laws, which were ideas in the mind of God. This mechanistic philosophy was revolutionary precisely because it rejected the animistic view of nature taken for granted in medieval Europe, as discussed in Chapter 1. Until the seventeenth century, university scholars and Christian theologians taught that the universe was alive, pervaded by the Spirit of God, the divine breath of life. All plants, animals and people had souls. The stars, the planets and the earth were living beings, guided by angelic intelligences.

Mechanistic science rejected these doctrines and expelled all souls from nature. The material world became literally inanimate, a soulless machine. Matter was purposeless and unconscious; the planets and stars were dead. In the entire physical universe, the only non-mechanical entities were human minds, which were immaterial, and part of a spiritual realm that included angels and God. No one could explain how minds related to the machinery of human bodies, but René Descartes speculated that they interacted in the pineal gland, the small pine-cone-shaped organ nestled between the right and left hemispheres near the centre of the brain.[21]

After some initial conflicts, most notably the trial of Galileo by the Roman Inquisition in 1633, science and Christianity were increasingly confined to separate realms by mutual consent. The practice of science was fairly free from religious interference, and religion fairly free from conflict with science, at least until the rise of militant atheism at the end of the eighteenth century. Science's domain was the material universe, including human bodies, animals, plants, stars and planets. Religion's realm was spiritual: God, angels, spirits and human souls. This more or less peaceful coexistence served the interests of both science and religion. Even in the late twentieth century Stephen Jay Gould still defended this arrangement as a 'sound position of general consensus'. He called it the doctrine of Non-overlapping Magisteria. The magisterium of science covers 'the empirical realm: what the Universe is made of (fact) and why does it work in this way (theory). The magisterium of religion extends over questions of ultimate meaning and moral value.'[22]

However, from around the time of the French Revolution (1789–99), militant materialists rejected this principle of dual magisteria, dismissing it as intellectually dishonest, or seeing it as a refuge for the feeble-minded. They recognised only one reality: the material world. The spiritual realm did not exist. Gods, angels and spirits were figments of the human imagination, and human minds were nothing but aspects or by-products of brain activity. There were no supernatural agencies that interfered with the mechanical course of nature. There was only one magisterium: the magisterium of science.

Atheist beliefs

The materialist philosophy achieved its dominance within institutional science in the second half of the nineteenth century, and was closely linked to the rise of atheism in Europe. Twenty-first-century atheists, like their predecessors, take the doctrines of materialism to be established scientific facts, not just assumptions.

When it was combined with the idea that the entire universe was like a machine running out of steam, according to the second law of thermodynamics, materialism led to the cheerless world-view expressed by the philosopher Bertrand Russell:

> That man is the product of causes which had no prevision of the end they were achieving; that his origin, his growth, his hopes and fears, his loves and beliefs, are but the outcome of accidental collisions of atoms; that no fire, no heroism, no intensity of thought and feeling, can preserve an individual life beyond the grave; that all the labours of the ages, all the devotion, all the inspiration, all the noonday brightness of human genius, are destined to extinction in the vast death of the solar system; and

that the whole temple of Man's achievement must inevitably be buried beneath the debris of a universe in ruins – all these things, if not quite beyond dispute, are yet so nearly certain, that no philosophy which rejects them can hope to stand. Only within the scaffolding of these truths, only on the firm foundation of unyielding despair, can the soul's habitation henceforth be built.[23]

How many scientists believe in these 'truths'? Some accept them without question. But many scientists have philosophies or religious faiths that make this 'scientific worldview' seem limited, at best a half-truth. In addition, within science itself, evolutionary cosmology, quantum physics and consciousness studies make the standard dogmas of science look old-fashioned.

It is obvious that science and technology have transformed the world. Science is brilliantly successful when applied to making machines, increasing agricultural yields and developing cures for diseases. Its prestige is immense. Since its beginnings in seventeenth-century Europe, mechanistic science has spread worldwide through European empires and European ideologies, like Marxism, socialism and free-market capitalism. It has touched the lives of billions of people through economic and technological development. The evangelists of science and technology have succeeded beyond the wildest dreams of the missionaries of Christianity. Never before has any system of ideas dominated all humanity. Yet despite these overwhelming successes, science still carries the ideological baggage inherited from its European past.

Science and technology are welcomed almost everywhere because of the obvious material benefits they bring, and the materialist philosophy is part of the package deal. However, religious beliefs and the pursuit of a scientific career can interact in surprising ways. As an Indian scientist wrote in the scientific journal *Nature* in 2009,

[In India] science is neither the ultimate form of knowledge nor a victim of scepticism . . . My observations as a research scientist of more than 30 years' standing suggest that most scientists in

India conspicuously evoke the mysterious powers of gods and goddesses to help them achieve success in professional matters such as publishing papers or gaining recognition.[24]

All over the world, scientists know that the doctrines of materialism are the rules of the game during working hours. Few professional scientists challenge them openly, at least before they retire or get a Nobel Prize. And in deference to the prestige of science, most educated people are prepared to go along with the orthodox creed in public, whatever their private opinions.

However, some scientists and intellectuals are deeply committed atheists, and the materialist philosophy is central to their belief system. A minority become missionaries, filled with evangelical zeal. They see themselves as old-style crusaders fighting for science and reason against the forces of superstition, religion and credulity. Several books putting forward this stark opposition were bestsellers in the 2000s, including Sam Harris's *The End of Faith: Religion, Terror, and the Future of Reason* (2004), Daniel Dennett's *Breaking the Spell* (2006), Christopher Hitchens's *God Is Not Great: How Religion Poisons Everything* (2007) and Richard Dawkins's *The God Delusion* (2006), which by 2010 had sold two million copies in English, and was translated into thirty-four other languages.[25] Until he retired in 2008, Dawkins was Professor of the Public Understanding of Science at the University of Oxford.

But few atheists believe in materialism alone. Most are also secular humanists, for whom a faith in God has been replaced by a faith in humanity. Humans approach a godlike omniscience through science. God does not affect the course of human history. Instead, humans have taken charge themselves, bringing about progress through reason, science, technology, education and social reform.

Mechanistic science in itself gives no reason to suppose that there is any point in life, or purpose in humanity, or that progress is inevitable. Instead it asserts that the universe is ultimately purposeless, and so is human life. A consistent atheism stripped of the humanist faith paints a bleak picture with little ground for hope, as Bertrand Russell made so clear. But secular humanism arose within a Judaeo-Christian culture and inherited from Christianity a belief in the unique importance of human life, together with a faith in future salvation. Secular humanism is in many ways a Christian heresy, in which man has replaced God.[26]

Secular humanism makes atheism palatable because it surrounds it with a reassuring faith in progress rather than provable facts. Instead of redemption by God, humans themselves will bring about human salvation through science, reason and social reform.[27]

Whether or not they share this faith in human progress, all materialists assume that science will eventually prove that their beliefs are true. But this too is a matter of faith.

Dogmas, beliefs and free enquiry

It is not anti-scientific to question established beliefs, but central to science itself. At the creative heart of science is a spirit of open-minded enquiry. Ideally, science is a process, not a position or a belief system. Innovative science happens when scientists feel free to ask new questions and build new theories.

In his influential book *The Structure of Scientific Revolutions* (1962), the historian of science Thomas Kuhn argued that in periods of 'normal' science, most scientists share a model of reality and a way of asking questions that he called a paradigm. The ruling paradigm defines what kinds of questions scientists can ask and how they can be answered. Normal science takes place within this framework and scientists usually explain away anything that does not fit. Anomalous facts accumulate until a crisis point is reached. Revolutionary changes happen when researchers adopt more inclusive frameworks of thought and practice, and are able to incorporate facts that were previously dismissed as anomalies. In due course the new paradigm becomes the basis of a new phase of normal science.[28]

Kuhn helped focus attention on the social aspect of science and reminded us that science is a collective activity. Scientists are subject to all the usual constraints of human social life, including peer-group pressure and the need to conform to the norms of the group. Kuhn's arguments were largely based on the history of science, but sociologists of science have taken his insights further by studying science as it is actually practised, looking at the ways that scientists build up networks of support, use resources and results to increase their power and influence, and compete for funding, prestige and recognition.

Bruno Latour's *Science in Action: How to Follow Scientists and Engineers Through Society* (1987) is one of the most influential studies in this tradition. Latour observed that scientists routinely make a distinction between knowledge and beliefs. Scientists within their professional group *know* about the phenomena covered by their field of science, while those outside the network have only distorted *beliefs*. When scientists think about people outside their groups, they often wonder how they can still be so irrational:

> [T]he picture of non-scientists drawn by scientists becomes bleak: a few minds discover what reality is, while the vast majority of people have irrational ideas or at least are prisoners of many social, cultural and psychological factors that make them stick obstinately to obsolete prejudices. The only redeeming aspect of this picture is that if it were only possible to *eliminate* all these factors that hold people prisoners of their prejudices, they would all, immediately and at no cost, become as sound-minded as the scientists, grasping the phenomena without further ado. In every one of us there is a scientist who is asleep, and who will not wake up until social and cultural conditions are pushed aside.[29]

For believers in the 'scientific worldview', all that is needed is to increase the public understanding of science through education and the media.

Since the nineteenth century, a belief in materialism has indeed been propagated with remarkable success: millions of people have been converted to this 'scientific' view, even though they know very little about science itself. They are, as it were, devotees of the Church of Science, or of scientism, of which scientists are the priests. This is how a prominent atheist layman, Ricky Gervais, expressed these attitudes in the *Wall Street Journal* in 2010, the same year that he was on the *Time* magazine list of the 100 most influential people in the world. Gervais is an entertainer, not a scientist or an original thinker, but he borrows the authority of science to support his atheism:

> Science seeks the truth. And it does not discriminate. For better or worse it finds things out. Science is humble. It knows what it knows and it knows what it doesn't know. It bases its conclusions and beliefs on hard evidence – evidence that is constantly updated and upgraded. It doesn't get offended when new facts come along. It embraces the body of knowledge. It doesn't hold onto medieval practices because they are tradition.[30]

Gervais's idealised view of science is hopelessly naïve in the context of the history and sociology of science. It portrays scientists as open-minded seekers of truth, not ordinary people competing for funds and prestige, constrained by peer-group pressures and hemmed in by prejudices and taboos. Yet naïve as it is, I take this ideal of free enquiry seriously. This book is an experiment in which I apply these ideals to science itself. By turning assumptions into questions I want to find out what science really knows and doesn't know. I look at the ten core doctrines of materialism in the light of hard evidence and recent discoveries. I assume that true scientists will not be offended when new facts come along, and that they will not hold onto the materialist worldview just because it's traditional.

I am doing this because the spirit of enquiry has continually liberated scientific thinking from unnecessary limitations, whether imposed from within or without. I am convinced that the sciences, for all their successes, are being stifled by outmoded beliefs.

1

Is Nature Mechanical?

Many people who have not studied science are baffled by scientists' insistence that animal and plants are machines, and that humans are robots too, controlled by computer-like brains with genetically programmed software. It seems more natural to assume that we are living organisms, and so are animals and plants. Organisms are self-organising; they form and maintain themselves, and have their own ends or goals. Machines, by contrast, are designed by an external mind; their parts are put together by external machine-makers and they have no purposes or ends of their own.

The starting point for modern science was the rejection of the older, organic view of the universe. The machine metaphor became central to scientific thinking, with very far-reaching consequences. In one way it was immensely liberating. New ways of thinking became possible that encouraged the invention of machines and the evolution of technology. In this chapter, I trace the history of this idea, and show what happens when we question it.

Before the seventeenth century, almost everyone took for granted that the universe was like an organism, and so was the earth. In classical, medieval and Renaissance Europe, nature was alive. Leonardo da Vinci (1452–1519), for example, made this idea explicit: 'We can say that the earth has a vegetative soul, and that its flesh is the land, its bones are the structure of the rocks . . . its breathing and its pulse are the ebb and flow of the sea.'[1] William Gilbert (1540–1603), a pioneer of the science of magnetism, was explicit in his organic philosophy of nature: 'We consider that the whole universe is animated, and that all the globes, all the stars, and also the noble earth have been governed since the beginning by their own appointed souls and have the motives of self-conservation.'[2]

Even Nicholas Copernicus, whose revolutionary theory of the movement of the heavens, published in 1543, placed the sun at the centre rather than the earth was no mechanist. His reasons for making this change were mystical as well as scientific. He thought a central position dignified the sun:

> Not unfittingly do some call it the light of the world, others the soul, still others the governor. Tremigistus calls it the visible God: Sophocles' Electra, the All-seer. And in fact does the sun, seated on his royal throne, guide his family of planets as they circle around him.[3]

Copernicus's revolution in cosmology was a powerful stimulus for the subsequent development of physics. But the shift to the mechanical theory of nature that began after 1600 was much more radical.

For centuries, there had already been mechanical *models* of some aspects of nature. For example, in Wells Cathedral, in the west of England, there is a still-functioning astronomical clock installed more than six hundred years ago. The clock's face shows the sun and moon revolving around the earth, against a background of stars. The movement of the sun indicates the time of day, and the inner circle of the clock depicts the moon, rotating once a month. To the delight of visitors, every quarter of an hour, models of jousting knights rush round chasing each other, while a model of a man bangs bells with his heels.

Astronomical clocks were first made in China and in the Arab world, and powered by water. Their construction began in Europe around 1300, but with a new kind of mechanism, operated by weights and escapements. All these early clocks took for granted that the earth was at the centre of the universe. They were useful models for telling the time and for predicting the phases of the moon; but no one thought that the universe was really like a clockwork mechanism.

A change from the metaphor of the organism to the metaphor of the machine-produced science as we know it: mechanical *models* of the universe were taken to represent the way the world *actually* worked. The movements of stars and planets were governed by impersonal mechanical principles, not by souls or spirits with their own lives and purposes.

In 1605, Johannes Kepler summarised his programme as follows: 'My aim is to show that the celestial machine is to be likened not to a divine organism but rather to clockwork . . . Moreover I show how this physical conception is to be presented through calculation and geometry.'[4] Galileo Galilei (1564–1642) agreed that 'inexorable, immutable' mathematical laws ruled everything.

The clock analogy was particularly persuasive because clocks work in a self-contained way. They are not pushing or pulling other objects. Likewise the universe performs its work by the regularity of its motions, and is the ultimate time-telling system. Mechanical clocks had a further metaphorical advantage: they were a good example of knowledge through construction, or knowing by doing. Someone who could construct a machine could reconstruct it. Mechanical knowledge was power.

The prestige of mechanistic science did not come primarily from its philosophical underpinnings but from its practical successes, especially in physics. Mathematical modelling typically involves extreme abstraction and simplification, which is easiest to realise with man-made machines or objects. Mathematical mechanics is impressively useful in dealing with relatively simple problems, such as the trajectories of cannon balls or rockets.

One paradigmatic example is billiard-ball physics, which gives a clear account of impacts and collisions of idealised billiard balls in a frictionless environment. Not only is the mathematics simplified, but billiard balls themselves are a very simplified system. The balls are made as round as possible and the table as flat as possible, and there are uniform rubber cushions at the sides of the table, unlike any natural environment. Think of a rock falling down a mountainside for comparison. Moreover, in the real world, billiard balls collide and bounce off each other in games, but the rules of the game and the skills and motives of the players are outside the scope of physics. The mathematical analysis of the balls' behaviour is an extreme abstraction.

From living organisms to biological machines

The vision of mechanical nature developed amid devastating religious wars in seventeenth-century Europe. Mathematical physics was attractive partly because it seemed to provide a way of transcending sectarian conflicts to reveal eternal truths. In their own eyes the pioneers of mechanistic science were finding a new way of understanding the relationship of nature to God, with humans adopting a God-like mathematical omniscience, rising above the limitations of human minds and bodies. As Galileo put it:

> When God produces the world, he produces a thoroughly mathematical structure that obeys the laws of number, geometrical figure and quantitative function. Nature is an embodied mathematical system.[5]

But there was a major problem. Most of our experience is not mathematical. We taste food, feel angry, enjoy the beauty of flowers, laugh at jokes. In order to assert the primacy of mathematics, Galileo and his successors had to distinguish between what they called 'primary qualities', which could be described mathematically, such as motion, size and weight, and 'secondary qualities', like colour and smell, which were subjective.[6] They took the real world to be objective, quantitative and mathematical. Personal experience in the lived world was subjective, the realm of opinion and illusion, outside the realm of science.

René Descartes (1596–1650) was the principal proponent of the mechanical or mechanistic philosophy of nature. It first came to him in a vision on 10 November 1619 when he was 'filled with enthusiasm and discovered the foundations of a marvellous science'.[7] He saw the entire universe as a mathematical system, and later envisaged vast vortices of swirling subtle matter, the aether, carrying around the planets in their orbits.

Descartes took the mechanical metaphor much further than Kepler or Galileo by extending it into the realm of life. He was fascinated by the sophisticated machinery of his age, such as clocks, looms and pumps. As a youth he designed mechanical models to simulate animal activity, such as a pheasant pursued by a spaniel. Just as Kepler projected the image of man-made machinery onto the cosmos, Descartes projected it onto animals. They, too, were like clockwork.[8] Activities like the beating of a dog's heart, its digestion and breathing were programmed mechanisms. The same principles applied to human bodies.

Descartes cut up living dogs in order to study their hearts, and reported his observations as if his readers might want to replicate them: 'If you slice off the pointed end of the heart of a live dog, and insert a finger into one of the cavities, you will feel unmistakably that every time the heart gets shorter it presses the finger, and every time it gets longer it stops pressing it.'[9]

He backed up his arguments with a thought experiment: first he imagined man-made automata that imitated the movements of animals, and then argued that if they were made well enough they would be indistinguishable from real animals:

> If any such machines had the organs and outward shapes of a monkey or of some other animal that lacks reason, we should have no way of knowing that they did not possess entirely the same nature as those animals.[10]

With arguments like these, Descartes laid the foundations of mechanistic biology and medicine that are still orthodox today. However, the machine theory of life was less readily accepted in the seventeenth and eighteenth centuries than the machine theory of the universe. Especially in England, the idea of animal-machines was considered eccentric.[11] Descartes' doctrine seemed to justify cruelty to animals, including vivisection, and it was said that the test of his followers was whether they would kick their dogs.[12]

As the philosopher Daniel Dennett summarised it, 'Descartes . . . held that animals were in fact just elaborate machines . . . It was only our non-mechanical, non-physical minds that make human beings (and only human beings) intelligent and conscious. This was actually a subtle view, most of which would readily be defended by zoologists today, but it was too revolutionary for Descartes' contemporaries.'[13]

We are so used to the machine theory of life that it is hard to appreciate what a radical break Descartes made. The prevailing theories of his time took for granted that living organisms were *organisms*, animate beings with their own souls. Souls gave organisms their purposes and powers of self-organisation. From the Middle Ages right up into the seventeenth century, the prevailing theory of life taught in the universities of Europe followed the Greek philosopher Aristotle and his leading Christian interpreter, Thomas Aquinas (*c.* 1225–74), according to whom the matter in plant or animal bodies was shaped by the organisms' souls. For Aquinas, the soul was the *form* of the body.[14] The soul acted like an invisible mould that shaped the plant or the animal as it grew and attracted it towards its mature form.[15]

The souls of animals and plants were natural, not supernatural. According to classical Greek and medieval philosophy, and also in William Gilbert's theory of magnetism, even magnets had souls.[16] The soul within and around them gave them their powers of attraction and repulsion. When a magnet was heated and lost its magnetic properties, it was as if the soul had left it, just as the soul left an animal body when it died. We now talk in terms of magnetic fields. In most respects fields have replaced the souls of classical and medieval philosophy.[17]

Before the mechanistic revolution, there were three levels of explanation: bodies, souls and spirits. Bodies and souls were part of nature. Spirits were non-material but interacted with embodied beings through their souls. The human spirit, or 'rational soul', according to Christian theology, was potentially open to the Spirit of God.[18]

After the mechanistic revolution, there were only two levels of explanation: bodies and spirits. Three layers were reduced to two by removing souls from nature, leaving only the human 'rational soul' or spirit. The abolition of souls also separated humanity from all other animals, which became inanimate machines. The 'rational soul' of man was like an immaterial ghost in the machinery of the human body.

How could the rational soul possibly interact with the brain? Descartes speculated that their interaction occurred in the pineal gland.[19] He thought of the soul as like a little man inside the pineal gland controlling the plumbing of the brain. He compared the nerves to water pipes, the cavities in the brain to storage tanks, the muscles to mechanical springs, and breathing to the movements of a clock. The organs of the body were like the automata in seventeenth-century water gardens, and the immaterial man within was like the fountain keeper:

> External objects, which by their mere presence stimulate [the body's] sense organs . . . are like visitors who enter the grottoes of these fountains and unwittingly cause the movements which take place before their eyes. For they cannot enter without stepping on certain tiles which are so arranged that if, for example, they approach a Diana who is bathing they will cause her to hide in the reeds. And finally, when a rational soul is present in this machine it will have its principal seat in the brain, and reside there like the fountain keeper who must be stationed at the tanks to which the fountain's pipes return if he wants to produce, or prevent, or change their movements in some way.[20]

The final step in the mechanistic revolution was to reduce two levels of explanation to one. Instead of a duality of matter and mind, there is only matter. This is the doctrine of materialism, which came to dominate scientific thinking in the second half of the nineteenth century. Nevertheless, despite their nominal materialism, most scientists remained dualists, and continued to use dualistic metaphors.

The little man, or homunculus, inside the brain remained a

common way of thinking about the relation of body and mind, but the metaphor moved with the times and adapted to new technologies. In the mid-twentieth century the homunculus was usually a telephone operator in the telephone exchange of the brain, and he saw projected images of the external world as if he were in a cinema, as in a book published in 1949 called *The Secret of Life: The Human Machine and How It Works*.[21] In an exhibit in 2010 at the Natural History Museum in London called 'How You Control Your Actions', you looked through a Perspex window in the forehead of a model man. Inside was a cockpit with banks of dials and controls, and two empty seats, presumably for you, the pilot, and your co-pilot in the other hemisphere. The ghosts in the machine were implicit rather than explicit, but obviously this was no explanation at all because the little men inside brains would themselves have to have little men inside their brains, and so on in an infinite regress.

If thinking of little men and women inside brains seems too naïve, then the brain itself is personified. Many popular articles and books on the nature of the mind say 'the brain perceives', or 'the brain decides', while at the same time arguing that the brain is just a machine, like a computer.[22] For example, the atheist philosopher Anthony Grayling thinks that 'brains secrete religious and superstitious belief' because they are 'hardwired' to do so:

> As a 'belief engine', the brain is always seeking to find meaning in the information that pours into it. Once it has constructed a belief, it rationalises it with explanations, almost always after the event. The brain thus becomes invested in the beliefs, and reinforces them by looking for supporting evidence while blinding itself to anything contrary.[23]

This sounds more like a description of a mind than a brain. Apart from begging the question of the relation of the mind to the brain, Grayling also begs the question of how his own brain escaped from this 'hardwired' tendency to blind itself to anything contrary to its beliefs. In practice, the mechanistic theory is only plausible because it smuggles non-mechanistic minds into human brains. Is a scientist operating mechanistically when he propounds a theory of materialism? Not in his own eyes. There is always a hidden reservation in his arguments: he is an exception to mechanistic determinism. He believes he is putting forward views that are true, not just doing what his brain makes him do.[24]

It seems impossible to be a consistent materialist. Materialism depends on a lingering dualism, more or less thinly disguised. In the realm of biology this dualism takes the form of personifying molecules, as I discuss below.

The God of mechanical nature

Although the machine theory of nature is now used to support materialism, for the founding fathers of modern science it supported the Christian religion, rather than subverted it.

Machines only make sense if they have designers. Robert Boyle, for example, saw the mechanical order of nature as evidence for God's design.[25] And Isaac Newton conceived of God in his own image as 'very well skilled in mechanics and geometry'.[26]

The better the world-machine functioned, the less necessary was God's ongoing activity. By the end of the eighteenth century, the celestial machinery was thought to work perfectly without any need for divine intervention. For many scientifically minded intellectuals, Christianity gave way to deism. A Supreme Being designed the world-machine, created it, set it in motion and left it to run automatically. This kind of God did not intervene in the world and there was no point in praying to him. In fact there was no point in any religious practice. Several Enlightenment philosophers, like Voltaire, combined deism with a rejection of the Christian religion.

Some defenders of Christianity agreed with the deists in accepting the assumptions of mechanistic science. The most famous proponent of mechanistic theology was William Paley, an Anglican priest. In his book *Natural Theology*, published in 1802, he argued that if someone were to find an object like a watch, he would be bound to conclude on examining it and observing its intricate design and precision that 'there must have existed, at some time and at some place or other, an artificer or artificers, who formed it for the purpose which we find it actually to answer, who comprehended its construction and designed its use'.[27] So it was with 'the works of nature' such as the eye. God was the designer.

In Britain in the nineteenth century, Anglican clergymen, most of whom emphasised the same points as Paley, wrote many popular books on natural history. For example, the Reverend Francis Morris wrote a popular, lavishly illustrated *History of British Butterflies* (1853), which served both as a field guide and a reminder of the beauty of nature. Morris believed that God had implanted in every human mind 'an instinctive general love of nature' through which young and old alike could enjoy the 'beautiful sights in which the benign Creator displays such infinite wisdom of Almighty skill'.[28]

This was the kind of natural theology that Darwin rejected in his theory of evolution by natural selection. By doing so, he undermined the machine theory of life itself, as I discuss below. But the controversy he stirred up is still with us, and its latest incarnation is Intelligent Design. Proponents of Intelligent Design point out the difficulty, if not impossibility, of explaining complex structures like the vertebrate eye or the bacterial flagellum in terms of a series of random genetic mutations and natural selection. They suggest that complex structures and organs show a creative integration of many different components because they were intelligently designed. They leave open the question of the designer,[29] but the obvious answer is God.

The problem with the design argument is that the metaphor of a designer presupposes an external mind. Humans design machines, buildings and works of art. In a similar way the God of mechanistic theology, or the Intelligent Designer, is supposed to have designed the details of living organisms.

Yet we are not forced to choose between chance and an external intelligence. There is another possibility. Living organisms may have an internal creativity, as we do ourselves. When we have a new idea or find a new way of doing something, we do not design the idea first, and then put it into our own minds. New ideas just happen, and no one knows how or why. Humans have an inherent creativity; and all living organisms may also have an inherent creativity that is expressed in larger or smaller ways. Machines require external designers; organisms do not.

Ironically, the belief in the divine design of plants and animals is not a traditional part of Christianity. It stems from seventeenth-century science. It contradicts the biblical picture of the creation of life in the first chapter of the Book of Genesis. Animals and plants were not portrayed as machines, but as self-reproducing organisms that arose from the earth and the seas, as in Genesis 1:11: 'And God said, Let the earth bring forth grass, the herb yielding seed, and the fruit trees yielding fruit after his kind, whose seed is in itself.' In Genesis 1: 24: 'God said, Let the earth bring forth the living creature after his kind, cattle and creeping thing and beast of the earth after his kind.' In theological language, these were acts of 'mediate' creation: God did not design or create these plants and animals directly. As an authoritative Roman Catholic Biblical Commentary expressed it, God created them indirectly 'through the agency of the mother earth'.[30]

When nature came to life again

Followers of the Enlightenment put their faith in mechanistic science, reason and human progress. 'Enlightened' ideas or values still have a major influence on our educational, social and political systems today. But from around 1780 to 1830 in the Romantic movement there was a widespread reaction against the Enlightenment faith, expressed mainly in the arts and literature. Romantics emphasised emotions and aesthetics, as opposed to reason. They saw nature as alive, rather than mechanical. The most explicit application of these ideas to science was by the German philosopher Friedrich von Schelling, whose book *Ideas for a Philosophy of Nature* (1797) portrayed nature as a dynamic interplay of opposed forces and polarities through which matter is 'brought to life'.[31]

A central feature of Romanticism was the rejection of mechani-cal metaphors and their replacement with imagery of nature as alive, organic, and in a process of gestation or development.[32] The first evolutionary theories arose in this context.

Some scientists, poets and philosophers linked their philoso-phy of living nature to a God who imbued Nature with life and left her to develop spontaneously, more like the God of Genesis than the designer God of mechanistic theology. Others proclaimed themselves atheists, like the English poet Percy Shelley (1792–1822), but they had no doubt about a living power in nature, which Shelley called the Soul of the universe, or the all-sufficing Power, or the Spirit of Nature. He was also a pioneering campaigner for vegetarianism because he valued animals as sentient beings.[33]

These different worldviews can be summarised as follows:

Worldview	God	Nature
Traditional Christian	Interactive	Living organism
Early mechanistic	Interactive	Machine
Enlightenment deism	Creator only	Machine
Romantic deism	Creator only	Living organism
Romantic atheism	No God	Living organism
Materialism	No God	Machine

The Romantic movement created an enduring split in Western culture. Among educated people, in the world of work, business and politics, nature is mechanistic, an inanimate source of natural resources, exploitable for economic development. Modern econo-mies are built on these foundations. On the other hand, children are often brought up in an animistic atmosphere of fairy tales, talking animals and magical transformations. The living world is celebrated in poems and songs and in works of art. Nature is most strongly identified with the countryside, as opposed to cities, and especially by unspoilt wilderness. Many urban people dream of moving to the country, or having a weekend home in rural surroundings. On Friday evenings, cities of the Western world are clogged with traffic as millions of people try to get back to nature in a car.

Our private relationship with nature presupposes that nature is alive. For a mechanistic scientist, or technocrat, or economist, or developer, nature is neuter and inanimate. It needs developing as part of human progress. But often the very same people have different attitudes in private. In Western Europe and North America, many people get rich by exploiting nature so that they can buy a place in the countryside to 'get away from it all'.

This division between public rationalism and private romanticism has been part of the Western way of life for generations, but is becoming increasingly unsustainable. Our economic activities are not separate from nature, but affect the entire planet. Our private and public lives are increasingly intertwined. This new consciousness is expressed through a revived public awareness of Gaia, Mother Earth. But goddesses were not far below the surface of scientific thought even in its most materialist forms.

The goddesses of evolution

One of the pioneers of evolutionary theory was Charles Darwin's grandfather, Erasmus Darwin, who wanted to increase the importance of nature and reduce the role of God.[34] The spontaneous evolution of plants and animals struck at the root of natural theology and the doctrine of God as designer. If new forms of life were brought forth by Nature herself, there was no need for God to design them. Erasmus Darwin suggested that God endued life or nature with an inherent creative capacity in the first place that was thereafter expressed without the need for divine guidance or intervention. In his book *Zoönomia* (1794), he asked rhetorically:

Would it be too bold to imagine that all warm-blooded animals have arisen from one living filament, which the great First Cause endued with animality, with the power of acquiring new parts, attended with new propensities, directed by irritations, sensations, volitions and associations, and thus possessing the faculty of continuing to improve by its own inherent activity, and of

delivering down these improvements by generation to its poster-
ity, world without end![35]

For Erasmus Darwin, living beings were self-improving, and the
results of the efforts of parents were inherited by their offspring.
Likewise, Jean-Baptiste Lamarck in his *Zoological Philosophy*
(1809) suggested that animals developed new habits in response to
their environment, and their adaptations were passed on to their
descendants. The giraffe, inhabiting arid regions of Africa,

> is obliged to browse on the leaves of trees and make constant
> efforts to reach them. From this habit long maintained in all its
> race, it has resulted that the animal's fore-legs have become longer
> than its hind legs, and its neck is lengthened to such a degree that
> the giraffe attains a height of six metres.[36]

In addition, a power inherent in life produced increasingly
complex organisms, moving them up a ladder of progress.
Lamarck attributed the origin of the power of life to 'the Supreme
Author', who created 'an order of things which gave existence
successively to all that we see'.[37] Like Erasmus Darwin, he was a
romantic deist. So was Robert Chambers, who popularised the
idea of progressive evolution in his best-selling *Vestiges of the
Natural History of Creation*, published anonymously in 1844. He
argued that everything in nature is progressing to a higher state as
a result of a God-given 'law of creation'.[38] His work was controversial both from a religious and scientific point of view but, like
Lamarck's theory, it was attractive to atheists because it removed
the need for a divine designer.

But Chambers, Lamarck and Erasmus Darwin not only undermined mechanistic theology, they also, perhaps unwittingly,
undermined the mechanistic theory of life. No inanimate
machinery contained within it a power of life, capacity for self-
improvement or creativity. Their theories of progressive evolution
demystified the creativity of God by mystifying evolution.

Charles Darwin and Alfred Russel Wallace's theory of evolution by natural selection (1858) attempted to demystify evolution. Natural selection was blind and impersonal, and required no divine agency. It weeded out organisms that were not fit to survive, and favoured those that were better adapted. The subtitle of Darwin's *On the Origin of Species* was *The Preservation of Favoured Races in the Struggle for Life*. The source of creativity was within animals and plants themselves: they varied spontaneously and adapted to new circumstances.

Darwin gave no explanation for this creative power. In effect, he rejected the designing God of mechanistic theology, and attributed all creativity to Nature, just as his grandfather had done. For Darwin, Nature herself gave rise to the Tree of Life. Through her prodigious fertility, her spontaneous variability and her powers of selection, she could do everything that Paley thought God did. But Nature was not an inanimate, mechanical system like the clockwork of celestial physics. She was Nature with a capital N. Darwin even apologised for his language: 'For brevity's sake I sometimes speak of natural selection as an intelligent power . . . I have, also, often personified the word Nature; for I have found it difficult to avoid this ambiguity.'[39]

Darwin advised his readers to ignore the implications of his turns of phrase. If, instead, we pay attention to their implications, Nature is the Mother from whose womb all life comes forth, and to whom all life returns. She is prodigiously fertile, but she is also cruel and terrible, the devourer of her own offspring. She is creative, but she is also destructive, like the Indian goddess Kali. For Darwin, natural selection was 'a power incessantly ready for action',[40] and natural selection worked by killing. The phrase 'Nature red in tooth and claw' was the poet Tennyson's rather than Darwin's, but sounds very like Kali, or the destructive Greek goddess Nemesis, or the vengeful Furies.

Charles Darwin, like his grandfather Erasmus and Lamarck, believed in the inheritance of habits. His books give many examples of offspring inheriting the adaptations of their parents.[41] The neo-Darwinian theory of evolution, which developed from the 1940s onwards, differed from Charles Darwin's theory in that it rejected the inheritance of acquired characteristics. Instead, organisms inherited genes from their parents, passing them on unaltered to their offspring, unless there were mutations, that is to say, random changes in the genes. The molecular biologist Jacques Monod summarised this theory in the title of his book, *Chance and Necessity* (1972).

These seemingly abstract principles are the hidden goddesses of neo-Darwinism. Chance is the goddess Fortuna, or Lady Luck. The turnings of her wheel confer both prosperity and misfortune. Fortuna is blind, and was often portrayed in classical statues with a veil or blindfold. In Monod's words, 'pure chance, absolutely free but blind, [is] at the very root of the stupendous edifice of evolution'.[42]

Shelley called Necessity the 'All-sufficing Power' and the 'Mother of the world'. She is also Fate or Destiny, who appears in classical European mythology as the Three Fates, who spin, allot and cut the thread of life, dispensing to mortals their destiny at birth. In neo-Darwinism, the thread of life is literal: helical DNA molecules in thread-like chromosomes dispense to mortals their destiny at birth.

Materialism is like an unconscious cult of the Great Mother. The word 'matter' itself comes from the same root as 'mother'; in Latin the equivalent words are *materia* and *mater*.[43] The Mother archetype takes many forms, as in Mother Nature, or Ecology, or even the Economy, which feeds and sustains us, working like a lactating breast on the basis of supply and demand. (The Greek root '*eco*' in both of these words means family or household.) Archetypes are more powerful when they are unconscious because they cannot be examined or discussed.

Life breaks out of mechanical metaphors

The theory of evolution destroyed the argument from mechanical design. A creator God could not have designed the machinery of animals and plants in the beginning if they evolved progressively through spontaneous variation and natural selection.

Living organisms, unlike machines, are themselves creative. Plants and animals vary spontaneously, respond to genetic changes and adapt to new challenges from the environment. Some vary more than others, and occasionally something really new appears. Creativity is inherent in living organisms, or works through them.

No machine starts from small beginnings, grows, forms new structures within itself and then reproduces itself. Yet plants and animals do this all the time. They can also regenerate after damage. To see them as machines propelled only by ordinary physics and chemistry is an act of faith; to insist that they are machines despite all appearances is dogmatic.

Within science itself, the machine theory of life was challenged continually throughout the eighteenth and nineteenth centuries by an alternative school of biology called vitalism. Vitalists thought that organisms were more than machines: they were truly vital or alive. Over and above the laws of physics and chemistry, organising principles shaped the forms of living organisms, gave them their purposive behaviour, and underlay the instincts and intelligence of animals. In 1844, the chemist Justus von Liebig made a typical statement of the vitalist position when he argued that although chemists could analyse and synthesise organic chemicals that occurred in living organisms, they would never be able to create an eye or a leaf. Besides the recognised physical forces, there was a further kind of cause that 'combines the elements in new forms so that they gain new qualities – forms and qualities which do not appear except in the organism'.[44]

In many ways, vitalism was a survival of the older worldview that living organisms were organised by souls. Vitalism was also in harmony with a romantic vision of living nature. Some vitalists, like the German embryologist Hans Driesch (1867–1941), deliberately used the language of souls to emphasise this continuity of thought. Driesch believed that a non-material organising principle gave plants and animals their forms and their goals. He called this organising principle *entelechy*, adopting a word that Aristotle had used for the aspect of the soul that has its end within itself (*en* = in, *telos* = purpose). Embryos, Driesch argued, behave in a purposive way; if their development is disrupted, they can still reach the form towards which they are developing. He showed by experiment that when sea-urchin embryos were cut in two, each half could give rise to a small but complete sea urchin, not half a sea urchin. Their entelechy attracted the developing embryos – and even separated parts of embryos – towards the form of the adult.

Vitalism was and still is the ultimate heresy within mechanistic biology. The orthodox view was clearly expressed by the biologist T. H. Huxley in 1867:

> Zoological physiology is the doctrine of the functions or actions of animals. It regards animal bodies as machines impelled by various forces, and performing a certain amount of work which can be expressed in terms of the ordinary forces of nature. The final object of physiology is to deduce the facts of morphology on the one hand, and those of ecology on the other, from the laws of the molecular forces of matter.[45]

In these words, Huxley foreshadowed the spectacular development of molecular biology since the 1960s, the most powerful effort ever made to reduce the phenomena of life to physical and chemical mechanisms. Francis Crick, who shared in a Nobel Prize for the discovery of the structure of DNA, made this agenda very explicit in his book *Of Molecules and Men* (1966). He denounced vitalism and affirmed his belief that 'the ultimate aim of the modern movement in biology is in fact to explain *all* biology in terms of physics and chemistry'.

The mechanistic approach is essentially reductionist: it tries to explain wholes in terms of their parts. That is why molecular biology has such a high status within the life sciences: molecules are some of the smallest components of living organisms, the point at which biology crosses over into chemistry. Hence molecular biology is at the leading edge of the attempt to explain the phenomena of life in terms of 'the laws of the molecular forces of matter'. In so far as biologists succeed in reducing organisms to the molecular level, they will then hand the baton to chemists and physicists, who will reduce the properties of molecules to those of atoms and subatomic particles.

Until the nineteenth century, most scientists thought that atoms were the solid, permanent, ultimate basis of matter. But in the twentieth century it became clear that atoms are made up of parts, with nuclei at the centre and electrons in orbitals around them. The nuclei themselves are made up of protons and neutrons, which in turn are composed of components called quarks, with three quarks each. When nuclei are split up in particle accelerators, like the Large Hadron Collider, at CERN, near Geneva, a host of further particles appears. Hundreds have been identified so far, and some physicists expect that with even larger particle accelerators, yet more will be found.

The bottom has dropped out of the atom, and a zoo of evanescent particles seems unlikely to explain the shape of an orchid flower, or the leaping of a salmon, or the flight of a flock of starlings. Reductionism no longer offers a solid atomic basis for the explanation of everything else. In any case, however many subatomic particles there may be, organisms are wholes, and reducing them to their parts by killing them and analysing their chemical constituents simply destroys what makes them organisms.

I was forced to think about the limitations of reductionism when I was a student at Cambridge. As part of the final-year biochemistry course, my class did an experiment on enzymes in rat livers. First, we each took a living rat and 'sacrificed' it over the sink, decapitating it with a guillotine, then we cut it open and removed its liver. We ground up the liver in a blender and centrifuged it, to remove unwanted fractions of the cellular debris. Then we purified the aqueous fraction to isolate the enzymes we wanted, and we put them in test tubes. Finally we added chemicals and studied the speeds at which chemical reactions took place. We learned something about enzymes, but nothing about how rats live and behave. In a corridor of the Biochemistry Department the bigger problem was summed up on a wall chart showing the chemical details of Human Metabolic Pathways; across the top someone had written in big blue letters, 'KNOW THYSELF'.

Attempts to explain organisms in terms of their chemical constituents are rather like trying to understand a computer by grinding it up and analysing its component elements, such as copper, germanium and silicon. Certainly it is possible to learn something about the computer in this way, namely what it is made of. But in this process of reduction, the structure and the programmed activity of the computer vanishes, and chemical analysis will never reveal the circuit diagrams; no amount of mathematical modelling of interactions between its atomic constituents will reveal the computer's programs or the purposes they fulfilled.

Mechanists expel purposive vital factors from living animals and plants, but then they reinvent them in molecular guises. One form of molecular vitalism is to treat the genes as purposive entities with goals and powers that go far beyond those of a mere chemical like DNA. The genes become molecular entelechies. In his book *The Selfish Gene*, Richard Dawkins endowed them with life and intelligence. Living molecules, rather than God, are the designers of the machinery of life:

> We are survival machines, but 'we' does not mean just people. It embraces all animals, plants, bacteria, and viruses . . . We are all survival machines for the same kind of replicator – molecules called DNA – but there are many different ways of making a living in the world, and the replicators have built a vast range of machines to exploit them. A monkey is a machine which preserves genes up trees; a fish a machine which preserves genes in the water.[46]

In Dawkins's words, 'DNA moves in mysterious ways.' The DNA molecules are not only intelligent, they are also selfish, ruthless and competitive, like 'successful Chicago gangsters'. The selfish genes 'create form', 'mould matter' and engage in 'evolutionary arms races'; they even 'aspire to immortality'. These genes are no longer mere molecules:

> Now they swarm in huge colonies, safe inside gigantic lumbering robots, sealed off from the outside world, communicating with it

by tortuous indirect routes, manipulating it by remote control. They are in you and me; they created us, body and mind; and their preservation is the ultimate rationale for our existence ... Now they go by the name of genes, and we are their survival machines.[47]

The persuasive power of Dawkins's rhetoric depended on anthropocentric language and his cartoon-like imagery. He admits that his selfish-gene imagery is more like science fiction than science,[48] but he justifies it as a 'powerful and illuminating' metaphor.[49]

The most popular use of a vitalistic metaphor in the name of mechanism is the 'genetic program'. Genetic programs are explicitly analogous to computer programs, which are intelligently designed by human minds to achieve particular purposes. Programs are purposive, intelligent and goal-directed. They are more like entelechies than mechanisms. The 'genetic program' implies that plants and animals are organised by purposive principles that are mind-like, or designed by minds. This is another way of smuggling intelligent designs into chemical genes.

If challenged, most biologists will admit that genes merely specify the sequence of amino acids in proteins, or are involved in the control of protein synthesis. They are not really programs; they are not selfish, they do not mould matter, or shape form, or aspire to immortality. A gene is not 'for' a characteristic like a fish's fin or the nest-building behaviour of a weaver bird. But molecular vitalism soon creeps back again. The mechanistic theory of life has degenerated into misleading metaphors and rhetoric.

To many people, especially gardeners and people who keep dogs, cats, horses or other animals, it is blindingly obvious that plants and animals are living organisms, not machines.

The philosophy of organism

Whereas the mechanistic and vitalist theories both date back to the seventeenth century, the philosophy of organism, also called the holistic or organismic approach, has been developing only since the 1920s. One of its proponents was the philosopher Alfred

North Whitehead (1861–1947); another was Jan Smuts, a South African statesman and scholar, whose book *Holism and Evolution* (1926) focused attention on 'the tendency of nature to form wholes that are greater than the sum of the parts through creative evolution'.[50] He saw holism as

> the ultimate synthetic, ordering, organising, regulative activity in the universe, which accounts for all the structural groupings and syntheses in it, from the atom and the physico-chemical structures, through the cell and organisms, through Mind in animals to Personality in man. The all-pervading and ever-increasing character of synthetic unity or wholeness in these structures leads to the concept of Holism as the fundamental activity underlying and co-ordinating all others, and to the view of the universe as a Holistic Universe.[51]

The holistic or organismic philosophy agrees with the mechanistic theory in affirming the unity of nature: the life of biological organisms is different in degree but not in kind from physical systems like molecules and crystals. Organicism agrees with vitalism in stressing that organisms have their organising principles within themselves; organisms are unities that cannot be reduced to the physics and chemistry of simpler systems.

The philosophy of organism in effect treats all nature as alive; in this respect it is an updated version of pre-mechanistic animism. Even atoms, molecules and crystals are organisms. As Smuts put it, 'Both matter and life consist, in the atom and the cell, of unit structures whose ordered grouping produces the natural wholes which we call bodies or organisms.'[52] Atoms are not inert particles of stuff, as in old-style atomism. Rather, as revealed by twentieth-century physics, they are structures of activity, patterns of energetic vibration within fields. In Whitehead's words, 'Biology is the study of the larger organisms, whereas physics is the study of the smaller organisms.'[53] In the light of modern cosmology, physics is also the study of very large organisms, like planets, solar systems, galaxies, and the entire universe.

The philosophy of organism points out that everywhere we look
in nature, at whatever level or scale, we find wholes that are made
up of parts that are themselves wholes at a lower level. This pattern
of organisation can be represented diagrammatically as in Figure
1.1. The smallest circles represent quarks, for example, within
protons, within atomic nuclei, within atoms, within molecules,
within crystals. Or the smallest circles represent organelles, in
cells, in tissues, in organs, in organisms, in societies of organisms,
in ecosystems. Or the smallest circles are planets, in solar systems,
in galaxies, in galactic clusters. Languages also show the same
kind of organisation, with phonemes in syllables, in words, in
phrases, in sentences.

FIGURE 1.1 A nested hierarchy of wholes or holons.

These organised systems are all *nested hierarchies*. At each level,
the whole includes the parts; they are literally within it. And at
each level the whole is more than the sum of the parts, with properties that cannot be predicted from the study of parts in isolation.
For example, the structure and meaning of this sentence could not
be worked out by a chemical analysis of the paper and the ink, or
deduced from the quantities of letters that make it up (five *a*s, one
b, five *c*s, two *d*s, etc.). Knowing the numbers of constituent parts
is not enough: the structure of the whole depends on the way they

are combined together in words, and on the relationships between the words.

Arthur Koestler proposed the term *holon* for wholes made up of parts that are themselves wholes:

> Every holon has a dual tendency to preserve and assert its individuality as a quasi-autonomous whole; and to function as an integrated part of an (existing or evolving) larger whole. This polarity between the Self-assertive and Integrative tendencies is inherent in the concept of hierarchic order.[54]

For such nested hierarchies of holons, Koestler proposed the term *holarchy*.

Another way of thinking about wholes is through 'systems theory', which speaks of 'a configuration of parts joined together by a web of relationships'.[55] Such wholes are also called 'complex systems', and are the subject of a number of mathematical models, variously called 'complex systems theory', 'complexity theory' or 'complexity science'.[56]

For a chemical example, think of benzene, a molecule with six carbon and six hydrogen atoms. Each of these atoms is a holon consisting of a nucleus with electrons around it. In the benzene molecule, the six carbon atoms are joined together in a six-sided ring, and electrons are shared between the atoms to create a vibrating cloud of electrons around the entire molecule. The patterns of vibration of the molecule affect the atoms within it, and since the electrons are electrically charged, the atoms are in a vibrating electromagnetic field. Benzene is a liquid at room temperature, but below 5.5°C it crystallises, and as it does so, the molecules stack themselves together in a regular three-dimensional pattern, called the lattice structure. This crystal lattice also vibrates in harmonic patterns,[57] creating vibrating electromagnetic fields, which affect the molecules within them. There is a nested hierarchy of levels of organisation, interacting through a nested hierarchy of vibrating fields.

In the course of evolution, new holons arise that did not exist before: for example, the first amino acid molecules, the first living cells, or the first flowers, or the first termite colonies. Since holons are wholes, they must arise by sudden jumps. New levels of organisation 'emerge' and their 'emergent properties' go beyond those of the parts that were there before. The same is true of new ideas, or new works of art.

The cosmos as a developing organism

The philosopher David Hume (1711–76) is perhaps best known today for his scepticism about religion. Yet he was equally sceptical about the mechanistic philosophy of nature. There was nothing in the universe to prove that it was more like a machine than an organism; the organisation we see in nature was more analogous to plants and animals than to machines. Hume was against the idea of a machine-designing God, and suggested instead that the world could have originated from something like a seed or an egg. In Hume's words, published posthumously in 1779,

> There are other parts of the universe (besides the machines of human invention) which bear still a greater resemblance to the fabric of the world, and which, therefore, afford a better conjecture concerning the universal origin of the system. These parts are animals and plants. The world plainly resembles more an animal or a vegetable, than it does a watch or a knitting-loom ... And does not a plant or an animal, which springs from vegetation or generation, bear a stronger resemblance to the world, than does any artificial machine, which arises from reason and design?[58]

Hume's argument was surprisingly prescient in the light of modern cosmology. Until the 1960s, most scientists still thought of the universe as a machine, and moreover as a machine that was running out of steam, heading for its final heat death. According to the second law of thermodynamics, promulgated in 1855, the universe would gradually lose the capacity to do work. It would eventually freeze in 'a state of universal rest and death', as William Thomson, later Lord Kelvin, put it.[59]

It was not until 1927 that Georges Lemaître, a cosmologist and Roman Catholic priest, advanced a scientific hypothesis like Hume's idea of the origin of the universe in an egg or seed. Lemaître suggested that the universe began with a 'creation-like event', which he described as 'the cosmic egg exploding at the moment of creation'.[60] Later called the Big Bang, this new cosmology echoed many archaic stories of origins, like the Orphic creation myth of the Cosmic Egg in ancient Greece, or the Indian myth of *Hiranyagarbha*, the primal Golden Egg.[61] Significantly, in all these myths the egg is both a primal unity and a primal polarity, since an egg is a unity composed of two parts, the yolk and the white, an apt symbol of the emergence of 'many' from 'one'.

Lemaître's theory predicted the expansion of the universe, and was supported by the discovery that galaxies outside our own are moving away from us with a speed proportional to their distance. In 1964, the discovery of a faint background glow everywhere in the universe, the cosmic microwave background radiation, revealed what seemed to be fossil light left over from the early universe, soon after the Big Bang. The evidence for an initial 'creation-like event' became overwhelming, and by 1966 the Big Bang theory became orthodox.

Cosmology now tells a story of a universe that began extremely small, less than the size of a pinhead, and very hot. It has been expanding ever since. As it grows, it cools down, and as it cools, new forms and structures appear within it: atomic nuclei and electrons, stars, galaxies, planets, molecules, crystals and biological life.

The machine metaphor has long outlived its usefulness, and holds back scientific thinking in physics, biology and medicine. Our growing, evolving universe is much more like an organism, and so is the earth, and so are oak trees, and so are dogs, and so are you.

What difference does it make?

Can you really think of yourself as a genetically programmed machine in a mechanical universe? Probably not. Probably even the most committed materialists cannot either. Most of us feel we are truly alive in a living world – at least at weekends. But through loyalty to the mechanistic worldview, mechanistic thinking takes over during working hours.

In recognising the life of nature, we can allow ourselves to recognise what we already know, that animals and plants are living organisms, with their own purposes and goals. Anyone who gardens or keeps pets knows this, and recognises that they have their own ways of responding creatively to their circumstances. But instead of dismissing our own observations and insights to conform to mechanistic dogma, we can pay attention to them and try to learn from them.

In relation to the living earth, we can see that the Gaia theory is not just an isolated poetic metaphor in an otherwise mechanical universe. The recognition of the earth as a living organism is a major step towards recognising the wider life of the cosmos. If the earth is a living organism, what about the sun and the solar system as a whole? If the solar system is a kind of organism, what about the galaxy? Cosmology already portrays the entire universe as a kind of growing super-organism, born through the hatching of the cosmic egg.

These differences in viewpoint do not immediately suggest a new range of technological products, and in that sense they may not be economically useful. But they make a big difference in healing the split created by the mechanistic theory – a split between our personal experiences of nature and the mechanical explanations that science gives us. And they help heal the split between the sciences and all traditional and indigenous cultures, none of which sees humans and animals as machines in a mechanical world.

Finally, dispelling the belief that the universe is an inanimate machine opens up many new questions, discussed in the following chapters.

Questions for materialists

Is the mechanistic worldview a testable scientific theory, or a metaphor?

If it is a metaphor, why is the machine metaphor better in every respect than the organism metaphor? If it is a scientific theory, how could it be tested or refuted?

Do you think that you yourself are nothing but a complex machine?

Have you been programmed to believe in materialism?

SUMMARY

The mechanistic theory is based on the metaphor of the machine. But it's only a metaphor. Living organisms provide better metaphors for organised systems at all levels of complexity, including molecules, plants and societies of animals, all of which are organised in a series of inclusive levels, in which the whole at each level is more than the sum of the parts, which are themselves wholes at a lower level. Even the most ardent defenders of the mechanistic theory smuggle purposive organising principles into living organisms in the form of selfish genes or genetic programs. In the light of the Big Bang theory, the entire universe is more like a growing, developing organism than a machine slowly running out of steam.

2

Is the Total Amount of Matter
and Energy Always the Same?

Every science student learns that the total amount of matter and energy is always the same. Matter and energy cannot be created or destroyed. The law of conservation of matter and energy is simple and reassuring: it guarantees fundamental permanence in an ever-changing world.

This law usually goes unquestioned. But it faces unprecedented challenges. As I discuss in this chapter, most physicists now believe that the universe contains large amounts of 'dark matter', whose nature and properties are literally obscure. Dark matter is currently thought to make up about 23 per cent of the mass and energy of the universe, whereas normal matter and energy make up only about four per cent. Worse still, most contemporary cosmologists think that the continuing expansion of the universe is driven by 'dark energy', whose nature is again obscure. According to the Standard Model of cosmology, dark energy currently accounts for about 73 per cent of the matter and energy of the universe.

How do dark matter and energy relate to regular matter and energy? And what is the zero-point energy field, also known as the quantum vacuum? Can any of this zero-point energy be tapped?

The law of conservation of matter and energy was formulated before these questions arose, and has no ready answer for them. It is based on philosophical and theological theories. Historically, it is rooted in the atomistic school of philosophy in ancient Greece. From the outset it was an assumption. In its modern form, it combines a series of 'laws' that have developed since the seventeenth century – the laws of conservation of matter, mass, motion, force and energy. In this chapter I look at the history of these ideas, and show how modern physics throws up questions that the old theories cannot answer. As faith in conservation comes into question, astonishing new possibilities open up in realms ranging from the generation of energy to human nutrition.

Matter, force and energy

Classical Newtonian physics was based on a fundamental distinction between matter and force. Matter was passive. Forces acted on matter causing changes. Material bodies either continued to exist in the same place for ever, or continued to move in a straight line perpetually, until they were acted on by forces that caused them to accelerate, or change direction or decelerate. Force was the active principle that caused change. Indeed, force or energy *was* causation. And because the cause must equal the effect, the total amount of force or energy must remain the same for logical reasons.

As the philosopher Immanuel Kant (1724–1804) made explicit, matter was inert and could only be experienced through its effects, and force was the *cause* of all these effects. In contrast with matter or bodies, forces and energies are not *things*: they are to do with processes in time. They are elusive. They breathe life, we might say poetically, into material nature and underlie all changes.

I begin with the history of the belief in the conservation of matter, which arose more than 2,500 years ago.

Eternal atoms

In ancient Greece, philosophers were preoccupied with the idea that behind the changing world of experience there was a changeless eternal reality, or an original unity. This conviction probably originated in mystical experiences, which appeared to reveal the existence of an ultimate reality or truth beyond space and time. The philosopher Parmenides tried to form an intellectual conception of an ultimate changeless being, and concluded that that being must be a changeless, undifferentiated sphere. There could be only one changeless thing, not many different things that change. But the world we experience contains many different things that change. Parmenides could only regard this as the result of illusion.

This conclusion was unacceptable to philosophers who came after him, for obvious reasons. They looked for more plausible theories of Absolute Being. Philosophers in the tradition of Pythagoras (c. 570–c. 495 BC) believed that eternal reality was made up of changeless mathematical truths. Plato and his followers thought in terms of transcendent Ideas or Forms beyond space and time. The atomist philosophers found another answer: Absolute Being is not a vast, undifferentiated, changeless sphere, but rather consists of many tiny, undifferentiated, changeless things – material atoms moving in the void. Thus the permanent atoms were the changeless basis of the changing phenomena of the world: matter was Absolute Being.[1] This philosophy of atomism or materialism, first propounded in the fifth century before Christ by Leucippus and Democritus,[2] was based on impressive feats of logical deduction. No one could see atoms or provide evidence for their existence, but it was a remarkably fruitful idea, and still exerts an enormous influence. Implicitly, the total amount of matter was always the same because the atoms were indestructible, by definition.

The atomists proposed that the movements and combinations of the atoms were governed by natural laws. There was no need for gods; neither were there any divine purposes in the universe. The human soul itself depended on combinations of atoms, and was extinguished at death; the atoms themselves continued for ever, entering into new permutations and combinations.

The main appeal of the atomist or materialist philosophy in pre-Christian Greece and Rome was its scepticism about the pantheon of gods and goddesses. Epicurus (341–270 BC), one of the most influential atomist philosophers, preached that materialism could liberate human beings from the fear of fickle gods and of divine retribution after death. He advocated a moderate form of hedonism, freed from these fears, teaching that happiness could best be achieved through simple pleasures and the company of friends.[3]

The Roman philosopher Lucretius (99–55 BC) popularised the Epicurean philosophy in his poem *De Rerum Natura*, 'On the Nature of Things'. He began by portraying Epicurus as the hero who crushed the monster of superstition and religion. He then explained everything mechanistically in terms of the purposeless motions and interactions of eternal atoms.

Atomistic materialism re-entered European thought from the late sixteenth century onwards largely through Lucretius's poem. It appealed to the founders of mechanistic science because it was mechanistic, not because it was anti-religious. The leading populariser of atomism was a French Roman Catholic priest, Pierre Gassendi (1592–1655), who tried to make the atomist doctrine compatible with Christianity. The founding fathers of mechanistic science followed his example by accepting God, the divine creation of the universe and the immortality of the soul as well as atoms of matter.

In effect, the seventeenth-century mechanistic theory of nature combined two Greek philosophies of eternity to produce a cosmic dualism: nature was made up of changeless atoms of matter in motion governed by immutable mathematical laws of nature that transcend space and time. But whereas for pre-Christian Greeks, like Democritus and Epicurus, atoms could be thought of as eternal, for the Christian founders of mechanistic science, they had to have been made by God in the first place.

Robert Boyle preferred to use the word 'corpuscle' because he wanted to avoid the atheistic implications of atomism and materialism. Boyle thought that in the creation of the universe God divided matter into a large number of small particles of different sizes and shapes, and isolated them from each other by setting them in motion in different ways.[4] After God had created them, the atoms just stayed the same. Isaac Newton agreed, and summarised his own views as follows:

> It seems probable to me, that God in the beginning formed matter in solid, massy, hard, impenetrable, movable particles . . . and that

these primitive particles being solids, are incomparably harder than any porous bodies compounded of them; even so very hard as to never wear or break in pieces; no ordinary power being able to divide what God himself made one in the first creation.[5]

In the late eighteenth century, atoms took on a more definite identity as the atoms of chemical elements. The pioneer of chemistry, Antoine Lavoisier (1743–94), took the law of conservation of mass or matter to mean that the total mass of all the products of a chemical reaction equalled the total mass of all the reactants. He defined an element as a basic substance that could not be further broken down by chemical methods, and was the first to recognise and name oxygen and hydrogen. Unfortunately Lavoisier was a tax collector as well as a chemist, and was guillotined at the height of the French Revolution. Soon afterwards, John Dalton (1766–1844) discovered that elements combine together in ratios of whole numbers, and he suggested that they involved combinations of chemical atoms, such as CO_2 and H_2O. The subsequent growth and enormous success of chemistry made atomism into an extremely fruitful theory.

The dissolution of solid matter

The more that atoms were investigated, the more apparent it became that they were not ultimate units of matter, made up of 'solid, massy, hard, impenetrable' particles, as Newton had imagined. Instead, they were structures of activity. From the 1920s onwards, quantum theory portrayed the constituent parts of atoms – electrons, nuclei and nuclear particles – as vibratory patterns of activity within fields. Like photons of light, they behave both as waves and as particles. As the philosopher of science Karl Popper expressed it, through modern physics, 'materialism transcended itself':[6]

Matter turns out to be highly packed energy, transformable into other forms of energy; and therefore something in the nature of a *process*, since it can be converted into other processes, such as light and, of course, motion and heat. Thus one may say that the results of modern physics suggest that we should give up *the idea of a substance or essence*. They suggest that there is no self-identical entity persisting during all changes in time . . . The universe now appears to be not a collection of things, but an interacting set of events or processes (as stressed especially by A. N. Whitehead).[7]

Meanwhile, according to the theory of quantum electrodynamics, brilliantly expounded by the physicist Richard Feynman, virtual particles, such as electrons, and photons, appear and disappear from the quantum vacuum field, also known as the zero-point field, that pervades the universe. Feynman called this theory the 'jewel of physics' because of its extremely accurate predictions, correct to many decimal places.

The price that is paid for this accuracy is the acceptance of invisible, unobservable particles and interactions, and of the mysterious quantum vacuum field. According to quantum electrodynamics, all electrical and magnetic forces are mediated by virtual photons that appear from the quantum vacuum field and then disappear into it again. When you look at a compass to find out where north is, the compass needle interacts with the earth's magnetic field through virtual photons. When you switch on a fan, its electric motor makes it go round because it is suddenly filled with virtual photons that exert forces. When you sit down, the chair supports your bottom because the chair and your bottom repel each other through a dense creation and destruction of virtual photons between them. When you get up, much of this activity in the vacuum field stops, and now great clouds of virtual photons appear between your feet and the floor, wherever you put your feet. All the molecules within your body, all your cell membranes, all your nerve impulses depend on virtual photons appearing and disappearing within the all-pervading vacuum field of nature. As the physicist Paul Davies put it, 'A vacuum is not inert and featureless, but alive with throbbing energy and vitality.'[8]

We have come a long way from a simple belief in atoms of matter as tiny solid objects that persist unchanged through time. According to current theories, matter itself is an energetic process, and mass depends on interactions with fields that pervade the vacuum.

Even mass, the quantitative measure of matter, turns out to be deeply mysterious. According to the Standard Model of particle physics, the mass of a particle like an electron or a proton is not inherent in the particle itself but depends on its interaction with a field called a Higgs field, named after one of the theoretical physicists who proposed it in 1964, Peter Higgs. Physicists think of this field as being like a universal pool of treacle that 'sticks' to otherwise massless particles travelling through it, conferring mass upon them.[9] Thus the mass of an electron, for example, arises through its interaction with the Higgs field, and this interaction depends on special Higgs particles, called Higgs bosons, which are hypothetical. There is no agreed prediction about their mass, and no Higgs boson has so far been detected, despite the expenditure of billions of euros to look for them in a gigantic particle accelerator, the Large Hadron Collider at CERN, near Geneva. Writers of popular science often refer to the Higgs boson as 'the God particle'. These elusive particles and fields have taken physics a long way from the Newtonian conception of matter as made up of 'solid, massy, hard, impenetrable, movable particles'.

The conservation of energy

What we now know as the law of conservation of energy did not emerge until the 1850s; indeed, the word 'energy' itself, though it came from a Greek root, was not in general use among scientists until the mid-nineteenth century. But right from the beginning of mechanistic science, there was a precursor of this law in the idea of conservation of motion or force. Like the conservation of matter, the conservation of motion or force was based on philosophical and theological arguments rather than on experimental observations.

For Descartes the original source of all matter and motion was God, and because God and his creation were immutable, the total quantity of matter and motion could not change. Individual particles could acquire or lose motion by colliding with other particles, but the total amount of motion was unaffected.[10] In the early nineteenth century, James Joule, who established the mechanical equivalent of heat, likewise made God the guarantor: '[T]he grand agents of nature are, by the Creator's fiat, indestructible; ... wherever mechanical force is expended, an exact equivalent of heat is always obtained.'[11] Michael Faraday was also convinced that God's powers could not be created or destroyed without some compensatory balance. He wrote, 'The highest law in physical science which our faculties permit us to perceive [is] the Conservation of Force.'[12]

In the first half of the nineteenth century, several different investigators arrived more or less independently at this conservation principle,[13] which became one of the great unifying principles of physics, combining ideas about kinetic energy, potential energy, heat, mechanical energy, chemical energy, light, electromagnetic energy and the energy of living organisms.[14] The forms of energy could change, but the total amount remained the same. The principle of conservation of energy was embodied in the first law of thermodynamics, which states that energy can be transformed from one form to another but cannot be created or destroyed.

As William Thomson, later Lord Kelvin, saw it, energy's fundamental status derived from its immutability and convertibility, and also from its unifying role in linking all physical phenomena in a web of energy transformations. He gave energy a theological sanction, and declared in 1852 that energy cannot be destroyed but only transformed 'as it is most certain that Creative Power alone can either call into existence or annihilate mechanical energy'.[15]

The ideas of conservation of matter and energy played an essential role in the development of the equations of physics. By definition, an equation demands that the total quantity of matter and energy before a change is equal to the total amount afterwards. In the 1960s, Richard Feynman expressed it as follows:

There is a fact, or if you wish, a *law*, governing all natural phenomena that are known to date. There is no known exception to this law; it is exact, so far as we know. The law is called the conservation of energy; it states that there is a certain quantity, which we call energy, that does not change in manifold changes which nature undergoes. That is a most abstract idea, because it is a mathematical principle; it says that there is a numerical quantity, which does not change when something happens. It is not a description of a mechanism, or anything concrete; it is just a strange fact that we can calculate some number, and when we finish watching nature go through her tricks and calculate the number again, it is the same.[16]

The principles of conservation of matter and energy were brought together by Albert Einstein in his famous equation $E=mc^2$, which shows the equivalence between mass (m), energy (E) and the velocity of light (c). For example, the amount of energy released as radiation in the explosion of an atomic bomb equals the amount of mass lost by the bomb, times the square of the velocity of light. However, the mass is not destroyed by being converted to radiant energy: the energy released by the bomb still has mass, and this mass is transferred to bodies that absorb the radiation. If the bomb loses one gram, and all its radiation is absorbed by other bodies, they collectively gain one gram. In effect, Einstein's equation meant that the conservation of matter became an aspect of the conservation of energy.

The equations of physics imply that satisfyingly precise relationships underlie all the transformations of nature. The conservation of matter and energy seems like a mathematical truth, even though matter is no longer solid, and mass depends on undetected Higgs particles. But the idea that the total amount of matter and energy is the same for ever runs into big problems in cosmology.

The appearance of matter from nowhere

The Big Bang theory, originally called the theory of the primeval atom, was first proposed in 1927 by Father Georges Lemaître. This theory became orthodox in the late 1960s.

The Big Bang theory means that all equations were violated in the primal singularity of the Big Bang. There was no conservation of matter and energy if the universe arose from nothing. As Terence McKenna expressed it, 'What orthodoxy teaches about time is that the universe sprang from utter nothingness in a single moment . . . It's almost as if science said, "Give me one free miracle, and from there the entire thing will proceed with a seamless, causal explanation."'[17] The one free miracle was the sudden appearance of all the matter and energy in the universe, with all the laws that govern it.

The creation of all matter and energy in the Beginning is presupposed by the Big Bang creation story, just as it was by René Descartes, Robert Boyle, Isaac Newton and other scientists who wanted to make physics compatible with an initial act of creation by God. Indeed in 1951, more than fifteen years before physicists generally accepted the Big Bang theory, Pope Pius XII welcomed it in an address to the Pontifical Academy of Sciences:

> Thus everything seems to indicate that the material universe had a mighty beginning in time, endowed as it was with vast reserves of energy, in virtue of which at first rapidly, and then ever more slowly, it evolved into its present state . . . In fact, it would seem that present-day science, with one sweeping step back across millions of centuries, has succeeded in bearing witness to that primordial *Fiat lux* uttered at the moment when, along with matter, there burst forth from nothing a sea of light and radiation.[18]

The Big Bang theory was initially controversial because some astronomers were suspicious of its theological implications; indeed, some opposed it precisely because the Pope approved of it. One British physicist suggested that the Big Bang theory was part of a conspiracy to shore up Christianity: 'The underlying motive is, of course, to bring in God as creator. It seems like the opportunity Christian theology has been waiting for ever since science began to depose religion from the minds of rational men in the seventeenth century.'[19] The astronomer Fred Hoyle condemned the Big Bang theory as a model built on Judaeo-Christian foundations,[20] and proposed an alternative. He argued that there was a process of continuous creation through which new matter and energy appeared within the universe as it expanded. The universe was eternal and infinite, and as the galaxies moved apart, new galaxies were created in the gaps between them. The universe was expanding yet remained in a steady state because of continuous creation, which took place as a result of the activity of a hypothetical C-field, or creation field, which both drove the steady expansion of the cosmos and generated new matter.

The original version of the steady state theory had to be abandoned because it predicted that new galaxies would be formed within the gaps between old ones, and hence young galaxies should be scattered all over the universe. By contrast, the Big Bang theory predicted that young galaxies would be formed relatively early in the history of the universe, and would therefore be found only far away, billions of light years in the past. In the early 1960s, evidence gathered by the British radio astronomer Martin Ryle showed that young galaxies were indeed distant, favouring the Big Bang theory. One of the theory's proponents, George Gamow, wrote a poem to celebrate:

> 'Your years of toil'
> Said Ryle to Hoyle
> 'Are wasted years, believe me,
> The Steady State
> Is out of date
> Unless my eyes deceive me.'[21]

Another discovery by a radio astronomer in 1963 seemed to provide further evidence for the Big Bang. Maartin Schmidt, a Dutch astronomer, was studying an extremely energetic radio source that he thought at first was a star in our own galaxy. But it turned out to have a high redshift: the radiation from it was much redder than would be expected if it were nearby. Objects far away have larger redshifts or, in other words, longer wavelengths of light than nearby objects because of the expansion of the universe.

Redshifts are produced by the Doppler effect: waves are stretched when their source is moving away, just as the sound waves from a siren get longer when a police car moves past; the tone drops. The further away galaxies are, the faster they are receding, and the redder they look. The high redshift of Schmidt's radio source suggested that this object was receding from us very fast. In fact it had the highest redshift ever detected, suggesting that it was over a billion light years away. This quasi-stellar radio object, or quasar, would therefore have to be an unprecedentedly brilliant galaxy, hundreds of times brighter than any yet known.

More quasars were soon discovered, and all of them had high redshifts and hence seemed very distant. If the universe were in a steady state, there should have been nearer quasars too, intense radio sources with small redshifts. But quasars seemed to lie in the most distant reaches of the universe.

The discovery in 1965 of the cosmic microwave background radiation, thought to be a kind of echo or afterglow of the Big Bang, seemed to settle the matter. Stephen Hawking described this discovery as 'the final nail in the coffin of the steady-state theory'. The Big Bang theory became the new orthodoxy. In the simplistic style of history favoured by many scientists, the Big Bang theory was the victor; the steady state was vanquished.

Dark matter

In the 1930s, Fritz Zwicky, a Swiss astrophysicist, studied the movements of galaxies within galactic clusters and realised that the clusters could not be held together by normal gravitation. Galaxies were attracting each other too strongly. The force holding them together seemed to be hundreds of times greater than a gravitational pull by visible matter could explain.[22]

Zwicky's results were ignored for decades, but were again taken seriously when it became apparent that the orbits of stars within galaxies could not be explained by the gravitational attraction of known kinds of matter. Too much force was acting upon the stars. Astronomers mapped the gravitational influences and found that apparent sources of gravitation did not correspond to the familiar disc-shaped structure of galaxies. Instead, there was a roughly spherical distribution of matter, which they called dark matter, stretching far beyond the fringes of the luminous galaxies, forming vast haloes extending into intergalactic space.[23]

Dark matter helps to explain the structures of galaxies and the relations between galaxies within clusters, but it does so at a heavy price: nobody knows what it is. Theories to explain it include vast numbers of unobserved black holes or other massive objects, or enormous quantities of undetected particles called WIMPS (weakly interacting massive particles).

A few physicists believe they can get rid of dark matter altogether by modifying the laws of gravitation instead.[24] If they are right, then the total amount of matter recognised by physics will drop dramatically.

Dark energy

In the mid-1990s, the problems for cosmologists worsened. Detailed observations of distant supernovas – exploding stars in faraway galaxies – showed that the expansion of the universe was speeding up. Gravitation ought to be slowing it down. So something else must account for accelerating growth. Physicists were forced to conclude that there must be an antigravity force, called dark energy, which they thought of in terms of a 'negative pressure' of empty space, or as an invisible field permeating the universe.

In 2010, only about four per cent of the universe was believed to be made up of familiar matter and energy such as atoms, stars, galaxies, gas clouds, planets and electromagnetic radiation.[25] Far from providing a satisfyingly complete explanation of the universe, modern physics suggests that we understand less than one twentieth of it. Moreover, some of the dark matter may be convertible into regular forms of energy. In 2010, observations of the centre of our galaxy showed that more gamma rays were being emitted than could be accounted for by known sources, leading some physicists to suggest that dark matter was being annihilated, giving rise to regular kinds of energy.[26]

In the light of modern cosmology, how can anyone be sure that the total amount of matter and energy has always been the same? As we have just seen, the standard kinds of matter and energy to which the conservation laws are supposed to apply are only a small fraction of the total amount of matter and energy. Most of the universe is composed of hypothetical dark matter and dark energy whose relationship to each other and to known kinds of matter and energy is mysterious. But the story becomes even more complicated. The amount of dark energy may be increasing.

Perpetual motion and the second law of thermodynamics

From the very beginning of modern science, there has been a denial of perpetual-motion machines as a matter of principle. Galileo proclaimed such machines could not exist, and so did most of the other founders of physics.[27] In the nineteenth century, Rudolf Clausius reformulated this prohibition in the second law of thermodynamics, which states that heat cannot flow spontaneously from a lower temperature to a higher temperature. In other words, heat does not flow 'uphill' unless aided by the expenditure of energy.[28]

Thermodynamics arose through the study of steam engines and was primarily concerned with heat, as the name 'thermodynamics' tells us. But the second law was soon generalised to cover other forms of energy as well. In general terms, this law gives a picture of energy flowing 'downhill', from a higher to a lower temperature, just as water powering a waterwheel flows downhill. In a watermill, the total amount of water remains the same although its ability to power the wheel is lost as it falls. Moreover, only some of the energy lost by the falling water as it powers the wheel is converted into useful work. Some is lost in friction and as heat; no machine is 100 per cent efficient.

From a thermodynamic point of view, machines are energy conversion devices, and only some of the energy can be converted into work. The rest is lost; it is dissipated into the surroundings as heat. This lost energy that cannot do work is measured in terms of *entropy*. In other words, entropy is a measure of the amount of energy that is not available for doing useful work in a machine or in any other thermodynamic process. More abstractly, the second law of thermodynamics states that spontaneous natural processes lead to an increase in entropy. Or, again, the entropy of a closed system always increases or remains constant: it does not decrease. This increase of entropy gives an arrow to time, and means that spontaneous processes are always running 'downhill' from a thermodynamic point of view.

When the second law of thermodynamics was generalised to the entire universe, it implied that the universe was like a machine running out of steam. Entropy would go on increasing until the universe froze for ever, the state described by William Thomson in 1852 as 'a state of universal rest and death'.[29] The ultimate heat death of the universe was the concept that underpinned Bertrand Russell's vision of 'the debris of a universe in ruins'.[30]

By contrast, evolutionary biology showed life evolving towards greater and greater complexity. The arrows of time in biology and physics were pointing in opposite directions. At first, this seeming disagreement was explained in terms of different time scales. Biological evolution was a temporary phenomenon on earth, but like the earth itself, it was ultimately doomed. But speculation about an ultimate heat death faded away when the Big Bang theory became orthodox in the 1960s. Cosmology itself became evolutionary: the universe began very small and very hot with little or no structure. As it grew and cooled, ever more complex forms of organisation came into being. Nevertheless some cosmological models suggested that this expanding, evolving universe would still come to an end: the gravity of the universe, magnified by the presence of dark matter, would make the universal expansion slow down, stop, and then give way to an ever-accelerating cosmic contraction, ending in a reverse of the Big Bang, the Big Crunch. Old-style cosmic pessimism based on the heat-death theory was replaced by a new kind.

In the late 1990s, the Big Crunch theory was replaced by a new vision of continued cosmic expansion powered by dark energy. In the current consensus, dark energy provides the motive force for the expansion of the universe, counteracting the gravitational pull that would otherwise cause it to contract. In most theoretical models, the density of dark energy in the universe is assumed to be constant; in other words, the amount of dark energy in a fixed physical volume remains the same. But the universe is expanding; its volume is increasing. Hence the total amount of dark energy in the universe is increasing.[31] The total amount of energy is *not* always the same. Far from running out of steam, the universe is now like a perpetual-motion machine, expanding because of dark energy, and creating more dark energy by expanding.

In the model currently favoured by most cosmologists, dark energy is uniform throughout the cosmos, but some models of dark energy propose that it arises from a 'quintessence' field that varies from place to place and time to time. The term 'quintessence', meaning 'fifth element', was borrowed from the ancient Greek term for the aether, which was thought to fill the universe. Quintessence interacts with matter and changes as the universe grows. It can also transform itself into new forms of hot matter or radiation, giving rise to new matter and energy.[32] Although the details differ, the creation of new matter and energy from the quintessence field recalls Hoyle's theory of the continuous creation of new matter and energy from a 'creation field'.

In this context, the laws of conservation of matter and energy seem less like ultimate cosmic principles and more like rules of accountancy that work reasonably well for most practical purposes in the realms of terrestrial physics and chemistry, where exotic possibilities like quintessence and the creation of dark energy can be ignored. In biology, the principle of conservation of energy is also a useful working assumption, but it may turn out to have papered over some fundamental cracks, as discussed below. Even in physical systems on earth, there may be energy conversion processes that have so far remained outside the scope of science, but which could be of practical importance in new technologies.

Alternative energy technologies

Scientific dogmas create taboos, with the result that entire areas of research and enquiry are excluded from mainstream science and from regular sources of funding. The result is 'fringe' science, kept beyond the pale of orthodoxy by automatic scepticism. As we have seen, one of the oldest and strongest taboos in science is against perpetual-motion machines, and this taboo extends to almost any kind of unconventional energy-generating device.

Many people claim to have made devices that produce 'free' energy using unconventional means. But they do not usually claim to have invented perpetual-motion machines. Instead they suggest that their devices are drawing on sources of energy that are usually untapped. Just as wind- and solar-power devices use freely available forms of energy, so some people claim to have made devices that tap into the zero-point energy or quantum-vacuum field, drawing on unlimited reserves of free power, while some claim to have found new ways of using electrical and magnetic forces. A search on the Internet for 'free energy devices' leads to a bewildering variety of claims and procedures. So does a search for 'over unity devices'. The term 'over unity' refers to the ability of a machine to produce more energy than is put into it. Sceptics claim that all these devices are impossible and/or fraudulent, and some promoters of 'free energy' devices may indeed be fraudulent. But can we be sure that they all are?

Do any of these devices really work? And if they do, why have they not already been taken up by entrepreneurs and marketed? One answer is that it is difficult to promote a device that appears to break the perpetual-motion taboo. As soon as a potential investor asks a scientific adviser, he is likely to be told that the device is impossible and would be a waste of money. But perhaps some of these devices really do work, and really can tap into new sources of energy.

This is an area in which offering a prize might provide the best way forward. In the history of science and technology, prizes have spurred several important innovations and also enabled inventors to attract publicity for their achievements. One of the first examples was the Longitude Prize, set up by the British government in 1714 for finding an accurate method to determine longitude at sea.[33] Another example is the Gossamer Condor, the first human-powered aircraft capable of sustained flight, which won the Kremer Prize in 1977. This prize was set up by a British industrialist, Harry Kremer, who offered £50,000 for the first group to fly a human-powered aircraft over a figure-of-eight course a mile long. The Gossamer Condor design was inspired by hang-gliders made of new lightweight materials and powered by an amateur cyclist. Its inventors went on to build the Gossamer Albatross, which flew twenty-two miles across the English Channel, winning the second Kremer Prize in 1979.

Current examples of incentivised challenges include the $10 million X Prizes, given by the X Prize Foundation 'to create radical breakthroughs for the benefit of humanity thereby inspiring the formation of new industries, jobs and the revitalisation of markets'.[34]

A prize for the most effective 'over unity' energy device might change the situation in energy research dramatically. In fair tests, conducted in an open-minded spirit of enquiry, some devices may indeed produce more energy than is put into them from conventional sources. Or perhaps the contest will reveal that no such devices exist, and no one will win the prize, giving scientific conservatives the pleasure of saying, 'I told you so.'

Energy conservation in living organisms

Until some of the theories of modern cosmology came along, energy conservation was uncontroversial in physics. But in biology, the situation was – and still is – less clear.

From the seventeenth century onwards, believers in the mechanistic philosophy asserted that living organisms were machines. Vitalists disagreed. This debate played an important part in the emergence of the idea of conservation of energy, especially in the work of Hermann von Helmholtz (1821–94). Although he is usually remembered as a leading German physicist, he was a medical doctor in the Prussian Army as a young man; his first researches were in physiology. When he was studying in Berlin, the vitalist doctrine held sway, teaching that living organisms depended on a 'life force' in addition to food, air and water. Helmholtz was an ardent believer in the mechanistic theory of life, and he made it his mission to rid biology of vitalism. At first he tried to refute the existence of vital force experimentally by studying the heat generated in the muscles of frogs' legs when they were stimulated to contract by electrical impulses. But it was hard to get clear-cut results so, having failed to prove it experimentally, he adopted a theoretical approach. He argued on philosophical grounds that perpetual-motion machines were impossible. Then, assuming that living organisms were indeed machines, he concluded that 'vital forces' did not exist. In 1847, when he was still only twenty-six, he published a memoir entitled 'On the Conservation of Force' that unified ideas about the conservation of force in living organisms, in physics and in machinery.[35]

Helmholtz's ideas were a major ingredient in the consensus on energy conservation that emerged in the 1850s. Living organisms were machines like everything else and obeyed the same laws, to which was now added the law of conservation of energy. From then onwards, this assumption was treated as an established fact. Indeed, as the mathematician Henri Poincaré pointed out, the very generality of the laws of conservation of matter and energy meant 'they are no longer capable of verification'.[36] Any evidence that went against them could be dismissed as flawed or fraudulent, or explained by invoking new forms of matter or energy hitherto unobserved.

Is the conservation of energy testable?

Helmholtz soon abandoned his attempts to prove the conservation of energy in frogs' legs. Other early attempts to measure the heat output compared with the energy released by respiration showed serious discrepancies, with 20 per cent more heat being produced than expected,[37] but the methods were crude and inaccurate. It was not until the 1890s that the energy balance of an animal was measured rigorously, long after the conservation laws had been assumed to apply to living organisms.

Max Rubner, working in Berlin, kept a dog in a specially constructed chamber, called a respiration calorimeter, for five weeks. The substance and energy content of its food were measured and its urine, faeces, carbon dioxide output and heat production were analysed. He found that the heat loss from the body agreed well with calculations of the amount of food material that was oxidised, with 99.7 per cent accuracy.[38] This was exactly what materialists wanted to hear, and the result was proclaimed to be the 'death knell of vitalism'.[39]

In the United States, in the early twentieth century, Wilbur Atwater and Francis Benedict carried out similar studies with human subjects using respiration calorimeters in order to 'demonstrate that man operated under the same laws that govern inanimate reactions'.[40] Like Rubner, the American researchers calculated the amount of energy that should have been released by the amount of food oxidised, and compared it with the energy output in terms of heat production plus work. The average for all their experiments gave a near-perfect agreement between measurements and calculations, just as they had expected.[41] This result was so convincing that it was unchallenged for more than seventy-five years.[42]

However, several other investigators were unable to replicate the expected results, and in a symposium on clinical calorimetry sponsored by the American Medical Association in 1921, a common complaint was that 'inexperienced operators were using the devices and obtaining inaccurate results'.[43] This comment highlights a general problem in scientific research. Results that agree with expectations are readily accepted, while those that do not are dismissed as flawed. And some experiments really are flawed – including some that give the expected results. Scientists, like most other people, accept evidence that agrees with their beliefs much more readily than evidence that contradicts them.

This is one reason why established orthodoxies in science remain established.

In the late 1970s, Paul Webb reinvestigated human energy balances in his laboratory in Ohio, with surprising results. The figures simply did not add up, especially when subjects were over-eating or under-eating. He looked again at the data from Atwater and Benedict's research, and found that some of their experiments showed serious discrepancies under conditions of vigorous exercise or under-eating. Atwater and Benedict's near-perfect results were arrived at by averaging data where too much or too little energy was consumed. Webb also found puzzling discrepancies in other previous studies. He concluded, 'The more careful the study, the more clearly there is evidence of energy not accounted for.'[44]

In Webb's own experiments, he took a careful tally of the food eaten over a three-week period, changes in body weight, heat and other forms of energy output, as well as measuring rates of oxygen consumption and carbon dioxide production. He found that more energy was being used than he could explain. He did not question the law of conservation of energy, but instead suggested that there was an as yet unrecognised and unmeasured kind of energy, which he called x. Taking all the studies together, x was on average 27 per cent of the total metabolic expenditure; in other words, more than a quarter of the energy was unaccounted for. Subsequent studies revealed further discrepancies in the energy balances of people who were gaining or losing weight, in pregnant women and in growing children.[45]

No one seemed worried about the problems revealed by Webb's research. The conservation of energy was not a question of evidence but an article of faith.

However, a modern-day vitalist could assert that there is a vital force at work in living organisms, over and above the standard forms of energy known to physics. A yogi could speak in terms of *prana*, or an acupuncturist in terms of *chi*. Do the available data rule out any kind of energy not yet known to physics? Is present-day nutrition science so precise that it can explain every detail of energetic activity in animals and people? The answer is 'no'. Careful, precise research might ultimately support the orthodox dogma, but at present it is an assumption, not a fact. Although most people do not realise it, there is a shocking possibility that living organisms draw upon forms of energy over and above those recognised by standard physics and chemistry.

One easy starting point for research would be to find out how some people and other animals seem to survive even though they eat very little food. It is well established that eating much less than usual can have beneficial effects. A reduced intake of calories, or 'caloric restriction', improves health, slows the ageing process and increases lifespan in a wide variety of species, including yeast, nematode worms, fruit flies, fish, rodents, dogs and people.

Inedia

A far greater challenge is presented by recurrent stories of people who seem to be able to live for months or years without eating. This phenomenon is known as *inedia* (Latin for 'fasting'). Of course such stories violate common sense: everybody knows that people, and animals, need food in order to stay alive.

I first heard of this phenomenon when my wife and I visited Jodhpur, in Rajasthan, India, in 1984. An Indian friend took us to visit a local holy woman called Satimata in the nearby village of Bala. We were told that when her husband died in 1943, when she was about forty years old, she wanted to immolate herself on his funeral pyre in the tradition of suttee, but she was prevented from doing so. Instead, she vowed never to eat again. When we met her, she was supposed to have lived for forty-three years without food or drink, and without producing faeces or urine. Yet she looked like a normal elderly village woman, apart from the fact that she was surrounded by devotees. While we were there she had a cold and had to blow her nose several times. So she seemed to be defying not only the law of conservation of energy but also the law of conservation of matter, generating mucus but taking in no food or water.

Of course I assumed that she must have been eating and drinking secretly. Yet her devotees were adamant that she was genuine. Some had known her for years, even lived with her, so had had the opportunity to see if she was eating behind the scenes. Either they were part of a conspiracy, or she was a very skilled deceiver. My scepticism was an immediate mental reflex. But when I met her, and talked to people who knew her, she did not strike me as a charlatan, but as a woman of sincere religious faith. I later found that she was not unique: other holy men and women in India were supposed to have lived without food for years. Some had been exposed as frauds, but others had been investigated by medical teams who found no evidence of secret eating.

In India, the explanation most commonly advanced for the ability to live without food is that the energy is derived from sunlight or from the breath, and in particular from *prana*, a life force in the breath. This is why some people who claim to live with little or no food call themselves 'breatharians'. Interestingly, the prana theory does not in itself challenge the principle of conservation of energy: it suggests that some people can derive all their energy from a source other than food.

In 2010, a team from the Indian Defence Institute of Physiology and Allied Sciences (DIPAS) investigated an eighty-three-year-old yogi called Prahlad Jani, who lived in the temple town of Anbaji in Gujarat. His devotees claimed that he had not eaten for seventy years. In the DIPAS study, he was kept for two weeks in a hospital under continuous observation and filmed on CCTV cameras. He had several baths and gargled, but the medical team confirmed that he ate and drank nothing, and passed no urine or faeces. A previous medical investigation in 2003 had given similar results. The director of DIPAS said, 'If a person starts fasting, there will be some changes in his metabolism but in his case we did not find any.'[46] This is an important point, because surviving a two-week fast is in itself not particularly impressive. Most people could do that, but there would be very noticeable physiological changes while they did so.

In the West, there have also been many claims that people can live for long periods without eating, including holy men and women like St Catherine of Siena (died 1380), St Lidwina (died 1433), who was said to have eaten nothing for twenty-eight years; the Blessed Nicholas von Flüe (died 1487), nineteen years; and the Venerable Domenica dal Paradiso (died 1553), twenty years. In the nineteenth century, two saintly women were said to have eaten nothing for twelve years except consecrated wafers in holy communion: Domenica Lazzari (died 1848) and Louise Lateau (died 1883).[47] In the nineteenth century there was also a wide-spread 'fasting girl' phenomenon in Europe and the United States. Some may have been anorexic, others were exposed as frauds; but there are some well-documented cases where girls lived for years without eating.

Herbert Thurston, a Jesuit scholar, documented this fascinating phenomenon in his classic study *The Physical Phenomena of Mysticism* (1952). He pointed out that not all cases of inedia occurred in particularly spiritual people. For example, a Scottish girl, Janet McLeod, seemed to survive without food for four years. She was investigated quite thoroughly and the case was reported in the *Philosophical Transactions of the Royal Society* in 1767. This young woman was seriously sick rather than saintly.

In the eighteenth century, Pope Benedict XIV asked the medical faculty of the University of Bologna to investigate cases of inedia. In their report, while fully recognising the likelihood of imposture, credulity and mal-observation, the doctors upheld 'the genuineness of certain well-attested examples of long abstinence from food though no supernatural causation could reasonably be supposed'.[48] As in the case of Janet McLeod, some of these cases seemed to result from illnesses.

The best-documented example in the twentieth century was the Bavarian mystic Therese Neumann (1898–1962). In 1922 she stopped eating solid food. On Fridays she had visions of the passion of Christ and, like some other Roman Catholic mystics, bled profusely from wounds on her hands and feet, known as stigmata. The astonishing nature of her prolonged fast as well as the stigmata attracted much public attention, and the Bishop of

Regensburg appointed a commission to investigate the case, headed by a distinguished doctor. Therese was closely observed for two weeks by a team of nursing sisters. Relieving each other by pairs, two of the four were continually on duty, never letting the girl out of their sight. Observation of her over a fortnight proved to the satisfaction of all unprejudiced persons that she did not during that period take either food or drink. What is even more striking, a pronounced loss of weight that occurred during the Friday ecstasies (owing to bleeding from her stigmata) was in each case made good during the two or three days that followed.[49]

But, as Thurston recognised, no amount of evidence would alter the opinions of committed sceptics, who declared her 'a vulgar imposter'. After considering many religious and non-religious cases, he concluded,

> We are forced to admit that quite a number of people, in whose case no miraculous intervention can be supposed, have lived for years upon a pittance of nourishing food which can be measured only by ounces; and upon this evidence we shall be forced to admit the justness of the conclusion of Pope Benedict XIV that mere continuation of life, while food and drink are withheld, cannot be safely assumed to be due to supernatural causes.[50]

If a pope and a leading Jesuit scholar favour a natural rather than supernatural explanation, what might it be? We will never find out by adopting a position of dogmatic scepticism and pretending that the phenomenon does not exist.

One starting point for research would be to find out where else in the world inedia occurs: it seems unlikely to be confined to India and the West. And if it occurs elsewhere, is it more common among females than males, as it seems to be in Europe?

What relationship does it bear to the physiology of hibernation in animals?

How is inedia related to 'caloric restriction'?

All these questions would greatly widen the scope of the science of nutrition, which is of increasing practical importance. About a billion people are classified as undernourished, while more than a billion are overweight or obese. There is a wide variety of dieting methods and no clear scientific consensus on what works best.

Including inedia within the field of science rather than keeping it beyond the pale might enable us to learn something important. By treating the laws of conservation and matter and energy as testable hypotheses rather than revealed truths, physiology and nutrition science would become more scientific, not less so.

Many people will confidently predict that all cases of inedia will be found to be fraudulent or to have some other conventional explanation. They may turn out to be right. If they are, the conventional assumptions will be strengthened by new evidence. But if they are wrong, we will learn something new that may raise bigger questions that go beyond the biological sciences. Are there new forms of energy that are not at present recognised by science? Or can the energy in the zero-point field, which *is* recognised by science, be tapped by living organisms?

What difference does it make?

The idea of matter as passive, and energy or force as the active principle of nature, is fundamental to science. It is also an ancient conception in religious traditions. The active principle is breath or spirit. Maybe there really is a free, creative spirit flowing through all nature, including the dark energy or quintessence through which the cosmos is growing. Our breath is part of this universal flow. We have mechanised the flow of energy through windmills, waterwheels, steam engines, motors and electric circuits, but outside man-made machines the flow is freer. Maybe the energy balances in galaxies, stars, planets, animals and plants are not always exact. Energy may not always be exactly conserved. And new matter and energy may arise from quintessence, more at some times and in some places than at others.

The flow of energy through living organisms may not depend only on the caloric content of food and the physiology of digestion and respiration. It may also depend on the way the organism is linked to a larger flow of energy in all nature. Terms like spirit, *prana* and *chi* may refer to a kind of energy that mechanistic science has missed out but which would show up quantitatively through discrepancies in calorimeter studies. If such a form of energy exists, how is it related to the principles of physics, including the zero-point field? Physiology may be seriously incomplete, and may have a lot to learn from non-mechanistic systems of healing, such as those of shamans, healers, and practitioners of yoga, ayurveda and acupuncture.

Meanwhile, modern physics reveals vast invisible reservoirs of dark matter and dark energy, and the quantum-vacuum field is full of energy, interacting with everything that happens. Maybe some of this energy can be tapped by new energy technologies, with huge economic and social consequences.

Questions for materialists

Is your belief in the conservation of matter and energy an assumption, or is it based on evidence? If so, what is the evidence?

Do you think that dark matter is conserved?

Can you accept that there may be a continuous creation of dark energy as the universe expands?

If there is a vast amount of energy in the quantum-vacuum field, do you think that we might be able to tap it?

SUMMARY

In the Big Bang all the matter and energy in the universe suddenly appeared from nowhere. Modern cosmology supposes that dark matter and dark energy now make up 96 per cent of reality. No one knows what dark matter and energy are, how they work or how they interact with familiar forms of matter and energy. The amount of dark energy seems to be increasing as the universe expands, and the 'quintessence field' may give rise to new matter and energy, more in some places than others. The evidence for energy conservation in living organisms is weak, and there are several anomalies, like the apparent ability of some people to live without food for long periods, that suggest the existence of new forms of energy. All quantum processes are supposed to be mediated through the quantum-vacuum field, also known as the zero-point field, which is not empty but full of energy and continually gives rise to virtual photons and particles of matter. Could this energy be tapped in new technologies?

3

Are the Laws of Nature Fixed?

Most scientists take it for granted that the laws of nature are fixed. They have always been the same as they are today, and will be for ever.

Obviously this is a theoretical assumption, not an empirical observation. On the basis of two or three hundred years of earth-bound research, how can we be sure that the laws were always the same and always will be, everywhere?

For most of science's history, eternal laws of nature made sense. Either the universe was eternal and had needed no God to create it, or it had been made by God and stayed the same thereafter, guaranteed by God's eternity. But in an evolutionary cosmos, does the theory of fixed laws make sense? Were all the laws of nature already present at the moment of the Big Bang, like a cosmic Napoleonic code? If everything else evolves, why don't the laws of nature evolve along with nature?

As soon as we begin to question them, eternal laws become problematic, for two main reasons. First, the very idea of a *law* of nature is anthropocentric. Only humans have laws. For the founders of modern science, the metaphor of law was appropriate because they thought of God as a kind of cosmic emperor whose writ ran everywhere, and whose omnipotence acted as a cosmic law-enforcement agency. The laws of nature were eternal ideas in the mind of a mathematical God. But for materialists there is no God and no transcendent mind in which these laws can be sustained. So where are they? And why do they still share the traditional attributes of God? Why are they universal, immutable and omnipotent, and why do they transcend space and time?

Some philosophers of science avoid these awkward questions by denying that scientific laws are transcendent, eternal realities; they argue instead that they are generalisations based on observable behaviour. But this amounts to an admission that the laws of nature evolve and may not be fixed for ever. In an evolutionary universe, nature evolves, so the generalisations that describe nature must evolve as well. There is no reason to assume that the laws that govern molecules, plants and brains were all present at the moment of the Big Bang, long before any of these systems existed.

Nevertheless, whatever some philosophers may say, eternal laws are deeply embedded in the thinking of most scientists. They are implicit in the scientific method. Any experiment should in principle be repeatable anywhere and at any time. Observations should be replicable. Why? Because the laws of nature are the same at all times and in all places.

In this chapter, I suggest an alternative to eternal laws: evolving habits. The regularities of nature do not depend on an eternal mind-like realm beyond space and time, but on a kind of memory inherent in nature.

A belief in eternal laws is itself deeply habitual, and often unconscious. To change a habit of thinking, the first step is to become aware of it. And this habit goes back a long way.

Eternal mathematics

The search by philosophers in ancient Greece for an eternal reality behind the changing world of appearances led to very different answers, as we saw in the previous chapter. The materialists thought that changeless atoms of matter were eternal, while Pythagoras and his followers believed that the entire universe, especially the heavens, was ordered according to eternal non-material principles of harmony. To understand mathematics was to link the human mind to the divine intelligence itself, which governed the creation with a transcendent perfection and order.[1] The Pythagoreans were more than philosophers: they formed mystical communities, shared property in common, treated men and women as equals, had vegetarian diets and believed in the transmigration of the soul. They thought that, through intellectual and moral discipline, the human mind could arrive at mathematical truths and begin to unravel the mysteries of the cosmos. They were convinced that the universe is governed by a regulating intelligence and that this same intelligence is reflected in the human mind.

Plato (428–348 BC) was strongly influenced by the Pythagoreans but went further. He generalised the notion of eternal mathematical truths to a wider vision of Forms or Ideas (Platonic Forms and Ideas are traditionally written with initial capitals), or archetypes or universals, including not only mathematics but the Forms of every object or quality, including horses, human beings, colours and goodness. These Forms or Ideas exist in an immaterial, transcendent realm, outside space and time. The cosmos is ordered by this realm that transcends it. The horses that we experience in the world are like shadows or reflections of the eternal essence of the horse, the horse Idea beyond space and time. All particular things in the world that we experience through our senses are reflections of transcendent Forms.

Plato famously compared the objects of sense experience to shadows in a cave experienced by prisoners, permanently chained so they can watch only the blank cave wall, with their backs to a fire. All they see are the shadows on the wall cast by things passing in front of the fire. In Plato's words,

> See what will naturally follow if the prisoners are released and disabused of their error. At first, when any of them is liberated and compelled suddenly to stand up and turn his neck round and walk and look towards the light, he will suffer sharp pains; the glare will distress him, and he will be unable to see the realities of which in his former state he had seen the shadows; and then conceive someone saying to him that what he saw before was an illusion but that now, when he is approaching nearer to being and his eye is turned towards more real existence, he has a clearer vision, what will be his reply? Will he not fancy that the shadows

which he formerly saw are truer than the objects which are now shown to him?[2]

Plato used the Greek word *nous* to signify the rational, immortal part of the soul through which the Forms could be known. As ancient philosophy progressed, the terms *logos* and *nous* were used to signify mind, reason, intellect, organising principle, word, speech, thought, wisdom and meaning. *Nous* was associated both with human reason and the universal intelligence.[3]

Many elements of Platonic philosophy were incorporated in Christian theology, and are implicit in the opening of St John's gospel, which, like the rest of the New Testament, was written in Greek. 'In the beginning was the Word.' 'Word' with a capital W is the translation of *logos*. Not long before St John's gospel was written, the word *logos* took on a new significance in the Jewish world when Philo of Alexandria (20 BC–AD 50) linked it to Jewish philosophy. Philo was a Greek-educated Jew, and the official representative of the Jewish community in Alexandria to the Roman emperor Caligula. He used *logos* to mean an intermediary divine being who bridged the gap between God and the material world. The Platonic Ideas were located in the *logos*, which Philo described as God's instrument in the creation of the universe. He compared God to a gardener forming the world according to the pattern of the *logos*.

In Europe from the fifteenth century onwards there was a revival of Platonism, which helped prepare the way for modern science. The founding fathers of modern science, Copernicus, Galileo, Descartes, Kepler and Newton, were all essentially Platonists or Pythagoreans. They thought that the business of science was to find the mathematical patterns underlying the natural world, the eternal mathematical Ideas that underlie all physical reality. As Galileo expressed it, Nature was a simple, orderly system that 'acts only through immutable laws which she never transgresses'. The universe was a 'book written in the mathematical language'.[4]

Most great physicists expressed similar ideas. For example, in the nineteenth century Heinrich Hertz, after whom the unit of frequency is named, expressed it as follows:

> One cannot escape the feeling that these mathematical formulae have an independent existence and an intelligence of their own, that they are wiser than we are, wiser even than their discoverers, that we get more out of them than was originally put into them.[5]

Albert Einstein's general theory of relativity was firmly in this tradition, and Arthur Eddington, who provided the first evidence in favour of the theory, concluded that it pointed to the idea that 'the stuff of the world is mind stuff . . . [T]he mind stuff is not spread out in space and time: these are part of the cyclic scheme ultimately derived out of it'.[6] The physicist James Jeans took a similarly Platonic view: '[T]he universe can be best pictured . . . as consisting of pure thought, the thought of what, for want of a wider word, we must describe as a mathematical thinker'.[7]

Quantum theory extended Platonism to the very heart of matter, which old-style atomists had regarded as hard, homogeneous stuff. In the words of Werner Heisenberg, one of the founders of quantum mechanics:

> [M]odern physics has definitely decided for Plato. For the smallest units of matter are not physical objects in the ordinary sense of the word: they are forms, structures, or – in Plato's sense – Ideas, which can be unambiguously spoken of only in the language of mathematics.[8]

The traditional assumption that the universe is governed by fixed laws and constant constants is almost unquestioned. This assumption has led to a baroque elaboration of theoretical speculation, including billions of extra universes as discussed below.

How constant are the 'fundamental constants'?

Some constants are considered to be more fundamental than others, including the velocity of light, c, the Universal Gravitational Constant, known to physicists as Big G, and the fine-structure constant, α, which is a measure of the strength of interaction between charged particles, such as electrons, and photons of light. Unlike the constants of mathematics, such as π, the values of the constants of nature cannot be calculated by mathematics alone: they depend on laboratory measurements. As the name implies, the physical constants are supposed to be changeless. They are believed to reflect an underlying constancy of nature. The standard assumption is that the laws and constants of nature are fixed for ever.

Are the constants really constant? The values given in handbooks of physics do in fact change from time to time. They are continually adjusted by international committees of experts known as metrologists. Old values are replaced by new 'best values' based on the latest data from laboratories around the world. Within their laboratories, metrologists strive for ever-greater precision. In so doing, they reject unexpected data on the grounds they must be errors. Then, after deviant measurements have been weeded out, they average the values obtained at different times, and the final value is then subjected to a series of corrections. Finally, in arriving at the latest 'best values', international committees of experts then select, adjust and average data from laboratories around the world.

Although the actual values change, most scientists take it for granted that the constants themselves are really constant; the variations in their values are simply a result of experimental errors. The latest values are the best, and previous values are forgotten. However, some physicists, notably Paul Dirac (1902–84), speculated that at least some of the fundamental constants might change with time. In particular, Dirac proposed that the Universal Gravitational Constant might decrease slightly as the universe expands. But Dirac was not challenging the idea of eternal mathematical laws: he was merely proposing that a mathematical law might govern the gradual variation of a constant.

What about the data? All of the published values of constants vary with time,[9] but here I will discuss only three of them: the Universal Gravitational Constant, the fine-structure constant and the speed of light.

The oldest of the constants, Newton's Universal Gravitational

Constant, Big G, is also the one that shows the largest variations. Towards the end of the twentieth century, as methods of measurement became more precise, the disparity in measurements of G by different laboratories increased, rather than decreased.[10] Between 1973 and 2010, the lowest value of G was 6.6659, and the highest 6.734, a 1.1 per cent difference (Figure 3.1). These published values are given to at least three places of decimals, and sometimes to five, with estimated errors of a few parts per million. Either this appearance of precision is illusory, or G really does change. The difference between recent high and low values is more than forty times greater than the estimated errors (expressed as standard deviations).[11]

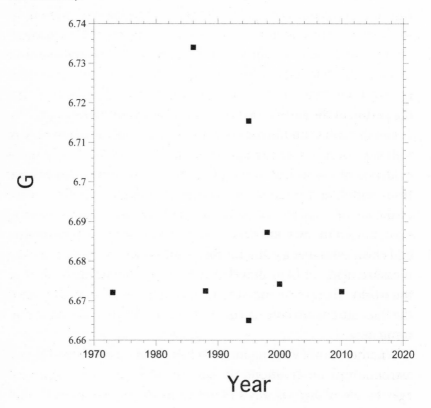

FIGURE 3.1 Values of G (x 10-[11]m[3]kg-[1]s-[2]) at different times between 1973 and 2010.[12]

What if G really does change? Maybe it does so because measurements are affected by changes in the earth's astronomical environment, as the earth moves around the sun and as the solar system moves within the galaxy. Or maybe there are inherent fluctuations in G. Such changes would never be noticed as long as measurements are averaged over time and across laboratories.

In 1998, the US National Institute of Standards and Technology published values of G taken on different days, rather than averaging them to iron out variations, revealing that there was a remarkable range: for example, on one day the value was 6.73, a few months later it was 6.64, 1.3 per cent lower.[13]

In 2002, a team led by Mikhail Gershteyn, of the Massachusetts Institute of Technology, published the first systematic attempt to study changes in G at different times of day and night. G was measured around the clock for seven months, using two independent methods. They found a clear daily rhythm, with maximum values of G 23.93 hours apart, correlating with the length of the sidereal day, the period of the earth's rotation in relation to the stars.

Gershteyn's team looked only for daily fluctuations, but G may well vary over longer time periods as well; there is already some evidence of an annual variation.[14] By comparing measurements from different locations, it should be possible to find more evidence of underlying patterns. Such measurements already exist, buried in the files of metrological laboratories. The simplest and cheapest starting point for this enquiry would be to collect the measurements of G at different times from laboratories all over the world. Then these measurements could be compared to see if the fluctuations are correlated.[15] If they are, we will discover something new.

Another way of looking for real changes in nature is to compare astronomical observations of galaxies and quasars of different ages to see if there is any difference in the light they emit that implies long-term changes in constants. The Australian astronomer John Webb has applied this approach to the fine structure constant, α.[16] Around the turn of the millennium, his team found that α was slightly smaller in distant parts of the sky, suggesting that it had changed over billions of years.[17] At first many physicists assumed that Webb's results must be due to errors, but by 2010 more data from different parts of the sky not only confirmed Webb's findings, but also gave new results that were quite unexpected. The variation in α depended on which way the telescopes were facing. The constant seemed to be larger on one side of the universe than the other. The variation of fundamental constants is now a matter of serious debate among physicists.[18] As Webb and his colleague John Barrow pointed out, 'If α is susceptible to change, other constants should vary as well, making the inner workings of nature more fickle than scientists ever suspected.'[19]

Finally, what about the speed of light, c? According to Einstein's theory of relativity, the speed of light in a vacuum is an absolute constant, and modern physics is based on this assumption.

Not surprisingly, early measurements of the speed of light varied considerably, but by 1927, the measured values had converged to 299,796 kilometres per second. At the time, the leading authority on the subject concluded, 'The present value of c is entirely satisfactory and can be considered more or less permanently established.'[20] However, all around the world from about 1928 to 1945, the speed of light dropped by about 20 kilometres per second.[21] The 'best' values found by leading investigators were in impressively close agreement with each other. Some scientists suggested that the data pointed to cyclic variations in the velocity of light.[22]

In the late 1940s the speed of light went up again by about 20 kilometres per second and a new consensus developed around the higher value. In 1972, the embarrassing possibility of variations in c was eliminated when the speed of light was fixed *by definition*. In addition, in 1983 the unit of distance, the metre, was redefined in terms of light. Therefore if any further changes in the speed of light happen, we will be blind to them because the length of the metre will change with the speed of light. (The metre is now defined as the length of the path travelled by light in a vacuum in 1/299,792,458 of a second.) The second is also defined in terms of light: it is the duration of 9,192,631,770 periods of vibration of the light given off by cesium 133 atoms in a particular state of excitation (technically defined as the transition between the two hyperfine levels of the ground state).

How can the drop in c between 1928 and 1945 be explained? This remarkable episode in the history of physics is now generally attributed to the psychology of metrologists. Brian Petley, a leading British metrologist, explained it thus:

> The tendency for experiments in a given epoch to agree with one another has been described by the delicate phrase 'intellectual phase locking'. Most metrologists are very conscious of the possible existence of such effects; indeed ever-helpful colleagues delight in pointing them out! Aside from the discovery of mistakes, the near completion of the experiment brings more frequent and stimulating discussion with interested colleagues and the preliminaries to writing up the work add a fresh perspective. All of these circumstances combine to prevent what was intended to be 'the final result' from being so in practice, and consequently the accusation that one is most likely to stop worrying about correction when the value is closest to other results is easy to make and difficult to refute.[23]

Existing theories of varying constants, like Paul Dirac's, assume that the changes are small, slow and systematic. Another possibility is that the constants oscillate within fairly narrow limits, or even vary chaotically. We are used to fluctuations in the weather and in human activities: newspapers and websites routinely report changes in the weather, stock-market indices, currency exchange rates and the price of gold. Maybe the constants fluctuate too, and perhaps one day scientific periodicals will carry regular news reports on their latest values.

The implications of varying constants would be enormous. The course of nature would no longer seem blandly uniform; there would be fluctuations at the heart of physical reality. If different constants varied at different rates, these changes would create differing qualities of time.

Multiple universes

According to the Anthropic Cosmological Principle, the fact that the 'laws' and 'constants' of nature are just right for human life on this planet requires an explanation. If these laws and constants had been even slightly different, carbon-based life would not exist. One response is to suggest that an Intelligent Designer fine-tuned the laws and constants of nature at the moment of the Big Bang so they were exactly right for the emergence of life and human beings. This is a modern version of deism. But an appeal to a divine mind, albeit of a remote, mathematical kind, is contrary to the atheistic spirit of much modern science. Instead many cosmologists prefer to think that there are innumerable actually existing universes besides our own, each with different laws and constants. In these 'multiverse' models the fact that we occupy a universe that is just right for us is explained very simply. This is the only universe that we can actually observe precisely because it is the only one right for us. No designer or divine mind was involved in making it so.[24]

The multiverse model is popular among cosmologists for two other reasons. First, models of an ultra-rapid period of inflation in the earliest stages of the Big Bang suggest that if this period of inflation could generate one universe, our own, it could also generate many others, and go on generating them.[25] This model, called eternal inflation, keeps creating 'pocket' universes, of which ours is merely one. The other theoretical reason for the popularity of the multiverse is superstring theory. This ten-dimensional theory and the related eleven-dimensional M-theory generate far too many possible solutions, which could correspond to different universes, as many as 10^{500}.[26]

Some theorists go even further. The cosmologist Max Tegmark proposes that any mathematically possible universe must exist somewhere: 'Complete mathematical democracy holds – mathematical existence and physical existence are equivalent, so that *all* mathematical structures exist physically as well.' There is no need to limit the mathematics to superstring theory or even any other existing mathematical system. Tegmark observes that his theory 'can be viewed as a form of radical Platonism'.[27]

In old-style Platonism, the mathematical laws were treated as if they were unique truths that transcended space and time, yet applied everywhere and always. By contrast, multiverse theories assume that particular laws and constants are built into each separate universe at the moment of its origin or Big Bang. They are somehow 'imprinted' in each universe. But how are they remembered? How does an individual universe 'know' what laws and constants are governing it, as opposed to the different laws and constants of other universes? As the cosmologist Martin Rees expressed it, 'The physical laws were themselves "laid down" in the Big Bang.'[28] But, he admitted, 'The mechanisms that might "imprint" the basic laws and constants in a new universe are obviously far beyond anything we understand.'[29]

Some physicists and cosmologists are unhappy with these speculations. A vast number of unobserved universes violates the canon of scientific testability. Multiverse supporters claim that mathematics itself, in the form of string and M-theories, provides evidence in favour of their speculations. But string and M-theories themselves, on which many of these speculations are based, are untestable. One critic, Peter Woit, called his book on the subject *Not Even Wrong*.[30] Even generic predictions that superstring theory shares with other theories, such as supersymmetry, have not fared well. The theoretical physicist Lee Smolin summarised the situation in 2006:

> Hundreds of careers and hundreds of millions of dollars have been spent in the last 30 years in the search for signs of grand unification, supersymmetry, and higher dimensions. Despite these efforts, no evidence for any of these hypotheses has turned up. A confirmation of any of these ideas, even if it could not be taken as a direct confirmation of string theory, would be the first indication that at least some parts of the package deal that string theory requires have taken us closer to, rather than further from, reality.[31]

Physicists who reject the multiverse theory have a variety of alternative suggestions. Some pin their faith on what they call a 'final theory', a unique mathematical formula that would predict every detail of our present universe, including all the so-called constants of nature. The uniqueness of the universe would then be a necessary consequence of mathematics.[32] This ultimate Platonic dream is far from coming true. But suppose that physicists really did one day discover The Formula. The next questions would be: where did it come from? And why did it exist in the first place? The answer would probably be a superformula. But where did *that* come from?

Another class of speculative theories suggests that the universe is part of a series of universes, the progeny of a previous one and the progenitor of the next. This is like the ancient Hindu philosophy of great cosmic cycles: the universe is born from the cosmic egg under the aegis of the god Brahma, its life and activity is sustained by Vishnu, and it is finally destroyed by Shiva. A new universe then arises. And so on. Or else the cycles are great breaths of Brahma, who breathes out a universe, breathes it in again, then breathes out another, and so on.

In modern cosmology, this ancient cyclic theory takes the form of the 'bouncing universe' model. After the Big Bang the universe expands for billions of years until its expansion slows down; it finally stops, and then begins to contract again under the force of gravity, finally collapsing in on itself in a Big Crunch. This in turn is the beginning of a new universe – a Big Bounce.[33]

One problem for this theory is that dark energy is currently thought to make the universe expand at an accelerating rate, so a collapse seems unlikely. To deal with this problem, the mathematician Roger Penrose has suggested that the exponential expansion of the universe will ultimately dilute everything so much that it will iron out all space-time features. Black holes will evaporate, stars and galaxies will disassemble and even elementary particles will decay into photons. Finally, the late universe will resemble the early universe except in size. Penrose magics this problem away by suggesting that, at these extremes, scale become irrelevant, and the late universe can become the early universe of the next in the series. Smolin has described this hypothesis as 'deliciously absurd, but just possibly true'.[34]

What all these theories have in common is a belief in the primacy of mathematics. Even if there are many universes besides our own, or a series of previous universes, what ultimately underlies these universes and sustains them? The answer is a mathematical formula transcending the universes it governs. In other words, this is a new, extravagant form of Platonism.

Evolutionary habits

The alternative to Platonism is the *evolution* of the regularities of nature. They are more like habits, and grow stronger through repetition. There is a kind of memory in nature: what happens now is influenced by what happened before.

Some habits run in very deep grooves and have been established for billions of years, like the habits of photons, protons and electrons, which existed before the first hydrogen atoms appeared about 370 million years after the Big Bang. As these first atoms arose they released the radiation presently observed as the cosmic microwave background.[35] Then, over billions of years, molecules, stars, galaxies, planets, crystals, plants and humanity appeared. Everything has evolved in time, even the chemical elements. At some point in the history of the universe the first carbon atoms, or iodine atoms, or gold atoms came into being.

The 'constants' associated with these atomic habits, like the fine-structure constant and the charge on the electron, are likewise very ancient. Among molecules, hydrogen, H_2, is probably the oldest; it precedes stars and is abundant in the galactic clouds from which new stars are formed. The 'laws' and 'constants' associated with these archaic patterns of organisation are so well established that they show little or no change today.

By contrast, some molecules are very new, like the hundreds of compounds made for the first time by synthetic chemists in the twenty-first century. Here the habits are still forming. So are new patterns of behaviour in animals, and new human skills.

In the late nineteenth century, the American philosopher Charles Sanders Peirce (1839–1914; pronounced 'purse') pointed out that fixed laws imposed upon the universe from the outset are inconsistent with an evolutionary philosophy. He was one of the first to propose that the 'laws of nature' are more like habits, and suggested that the tendency to form habits grows spontaneously: 'There were slight tendencies to obey rules that had been followed, and these tendencies were rules which were more and more obeyed by their own action.'[36] Peirce considered that 'the law of habit is the law of mind' and he saw the growing cosmos as alive. 'Matter is merely mind deadened by the development of habit to the point where the breaking up of these habits is very difficult.'[37]

The German philosopher Friedrich Nietzsche (1844–1900), writing around the same time, went so far as to suggest that the 'laws of nature' underwent natural selection:

> At the beginning of things we may have to assume, as the most general form of existence, a world which was not yet mechanical, which was outside all mechanical laws, although having access to them. Thus the origin of the mechanical world would be a lawless game which would ultimately acquire such consistency as the organic laws seem to have now ... All our mechanical laws would not be eternal, but evolved, and would have survived innumerable alternative mechanical laws.[38]

The philosopher and psychologist William James (1842–1910) wrote in a similar vein to Peirce:

> If ... one takes the theory of evolution radically, one ought to apply it not only to the rock-strata, the animals and plants, but to the stars, to the chemical elements, and to the laws of nature. There must have been a far-off antiquity, one is then tempted to suppose, when things were really chaotic. Little by little, out of all the haphazard possibilities of that time, a few connected things and habits arose, and the rudiments of regular performance began.[39]

*   *   *

Likewise, Alfred North Whitehead suggested, 'Time is differenti-
ated from space by the acts of inheriting patterns from the past.'
This inheritance of pattern meant that habits built up. Whitehead
said, 'People make the mistake of talking about "natural laws". There
are no natural laws. There are only temporary habits of nature.'[40]

These philosophers were far ahead of their time. They thought
the entire universe was evolutionary. But their physicist contemporaries still believed in an eternal universe, made of permanent
matter and energy governed by changeless laws, devolving towards
a heat death, according to the second law of thermodynamics. The
Big Bang theory became orthodox only in the 1960s. As Peirce,
James and Whitehead saw so clearly, evolutionary cosmology
implies the evolution of habits.

Morphic resonance

My own hypothesis is that the formation of habits depends on a
process called morphic resonance.[41] Similar patterns of activity
resonate across time and space with subsequent patterns. This
hypothesis applies to all self-organising systems, including atoms,
molecules, crystals, cells, plants, animals and animal societies. All
draw upon a collective memory and in turn contribute to it.

A growing crystal of copper sulphate, for example, is in resonance
with countless previous crystals of copper sulphate, and follows the
same habits of crystal organisation, the same lattice structure. A
growing oak seedling follows the habits of growth and development
of previous oaks. When an orb-web spider starts spinning its web, it
follows the habits of countless ancestors, resonating with them
directly across space and time. The more people who learn a new
skill, such as snowboarding, the easier will it be for others to learn it
because of morphic resonance from previous snowboarders.

In summary, this hypothesis proposes that:

1. Self-organising systems including molecules, cells, tissues,
   organs, organisms, societies and minds are made up of
   nested hierarchies or holarchies of holons or morphic units

(Figure 1.1). At each level the whole is more than the sum of the parts, and these parts themselves are wholes made up of parts.

2. The wholeness of each level depends on an organising field, called a morphic field. This field is within and around the system it organises, and is a vibratory pattern of activity that interacts with electromagnetic and quantum fields of the system. The generic name 'morphic field' includes

   (a) Morphogenetic fields that shape the development of plants and animals.

   (b) Behavioural and perceptual fields that organise the movements, fixed-action patterns and instincts of animals.

   (c) Social fields that link together and co-ordinate the behaviour of social groups.

   (d) Mental fields that underlie mental activities and shape the habits of minds.

3. Morphic fields contain attractors (goals), and chreodes (habitual pathways towards those goals) that guide a system towards its end state, and maintain its integrity, stabilising it against disruptions (see Chapter 5).

4. Morphic fields are shaped by morphic resonance from all similar past systems, and thus contain a cumulative collective memory. Morphic resonance depends on similarity, and is not attenuated by distance in space or time. Morphic fields are local, within and around the systems they organise, but morphic resonance is non-local.

5. Morphic resonance involves a transfer of form or in-*form*-ation rather than a transfer of energy.

6. Morphic fields are fields of probability, like quantum fields, and they work by imposing patterns on otherwise random events in the systems under their influence.

7. All self-organising systems are influenced by self-resonance from their own past, which plays an essential role in maintaining a holon's identity and continuity.

This hypothesis leaves open the question of how morphic resonance actually works. There are several suggestions. One is that the transfer of information occurs through the 'implicate order', as proposed by the physicist David Bohm.[42] The implicate or enfolded order gives rise to the world we can observe, the explicate order, in which things are located in space and time. In the implicate order, according to Bohm, 'everything is enfolded into everything'.[43] Or resonance may pass through the quantum-vacuum field, also known as the zero-point energy field, which mediates all quantum and electromagnetic processes (see Chapter 2).[44] Or similar systems might be connected through hidden extra dimensions, as in string theory and M-theory.[45] Or maybe it depends on new kinds of physics as yet unthought of.

This hypothesis is eminently testable, and evidence from many fields of enquiry already supports it. I discuss tests in the realm of biological development and animal behaviour in Chapter 6, and in human learning in Chapter 7.

Habits of crystallisation

The hypothesis of morphic resonance predicts that when chemists make a new compound for the first time, it might be difficult to obtain crystals of this compound because a morphic field for this crystal-form does not yet exist. When the crystals appear for the first time, a new pattern of organisation comes into being. The second time the compound crystallises, there will be an influence from the first crystals by morphic resonance all over the world. The third time, there will be an influence from the first and the second crystals, and so on. This influence builds up cumulatively. A new habit develops. The more a compound crystallises, the easier its crystals should form.

In fact, when chemists synthesise new chemicals, they often have great difficulty in getting them to crystallise. Sometimes it takes many years before crystals first appear. For example, turanose, a kind of sugar, was considered to be a liquid for decades, until it first crystallised in the 1920s. Thereafter it formed crystals all over the world.[46] In many other cases, new compounds crystallised with greater and greater ease as time went on.

Even more striking are cases in which one kind of crystal appeared, and was then replaced by another. Xylitol, a sugar alcohol used as a sweetener in chewing gum, was first prepared in 1891 and was considered to be a liquid until 1942, when crystals first appeared. The melting point of these crystals was 61°C. After a few years another crystal form appeared, with a melting point of 94°C, and thereafter the first kind of crystals did not show up again.[47]

Crystals of the same compound that exist in different forms are called polymorphs. Sometimes they coexist, like calcite and aragonite, both crystalline forms of calcium carbonate, or graphite and diamond, both crystalline forms of carbon. But sometimes, as in the case of xylitol, the appearance of a new polymorph can displace an old one. This principle is illustrated in the following account, taken from a textbook on crystallography, of the spontaneous and unexpected appearance of a new type of crystal in a factory:

> [A] company was operating a factory which grew large single crystals of ethylene diamine tartrate from solution in water. From this plant it shipped the crystals many miles to another which cut and polished them for industrial use. A year after the factory opened, the crystals in the growing tanks began to grow badly; crystals of something else adhered to them – something which grew even more rapidly. The affliction soon spread to the other factory: the cut and polished crystals acquired the malady on their surfaces . . . The wanted material was *anhydrous* ethylene diamine tartrate, and the unwanted material turned out to be the *monohydrate* of that substance. During three years of research and development, and another year of manufacture, no seed of the monohydrate had formed. After that they seemed to be everywhere.[48]

These authors suggest that on other planets, types of crystal that are common on earth may not yet have appeared, and add: 'Perhaps in our own world many other possible solid species are still unknown, not because their ingredients are lacking, but simply because suitable seeds have not yet put in an appearance.'[49]

The replacement of one polymorph by another is a recurrent problem in the pharmaceutical industry. For example, the antibiotic ampicillin was first crystallised as a monohydrate, with one molecule of water of crystallisation per ampicillin molecule. In the 1960s it started to crystallise as a trihydrate, with a different crystal form, and despite persistent efforts, the monohydrate could not be made again.[50]

Ritonavir, an AIDS drug made by Abbott Laboratories, was introduced in 1996. The drug had been on the market for eighteen months when chemical engineers found a previously unknown polymorph. No one knew what had caused the change, and the Abbott team could not stop the new polymorph forming. Within a few days of its discovery, it was dominating the production lines. Although the two polymorphs had the same chemical formula, the second form was only half as soluble as the first, so patients taking the normal prescribed doses would not absorb enough of the drug. Abbott had to withdraw Ritonavir from the market and launched a crash programme to get their original polymorph back. They eventually succeeded, but could not make it reliably: they kept getting mixtures of the two forms. The company finally decided to reformulate the drug as a capsule containing the drug in solution. It spent hundreds of millions of dollars on this process and also lost an estimated $250 million dollars in sales the year the drug was withdrawn.[51]

The inability of chemists to achieve control of crystallisation is a serious challenge. 'The loss of control is indeed disturbing, and might even call into question the criterion of reproducibility as a condition for acceptance of a phenomenon as being worthy of scientific enquiry,' wrote Joel Bernstein, in his book *Polymorphism in Molecular Crystals*.[52] The emergence of new polymorphs makes it clear that chemistry is not timeless. It is historical and evolutionary, like biology. What happens now depends on what has happened before.

One possible explanation for the disappearance of polymorphs is that the new forms are more stable thermodynamically, and hence appear in preference to the older forms; in competition with each other, the new forms win. Before the new forms existed, there was no competition; after they had come into being, they turned up in laboratories all over the world, and the older forms disappeared.

There is no doubt that small fragments of previous crystals can act as 'seeds' or 'nuclei' facilitating the process of crystallisation from a supersaturated solution. That is why chemists assume that the spread of new crystallisation processes depends on the transfer of nuclei from laboratory to laboratory, like a kind of infection. One favourite story in the folklore of chemistry is that migrant scientists carry these seeds around the world from lab to lab. Some chemists are said to have beards that 'harbour nuclei for almost any crystallisation process', in the words of a professor of chemical engineering at Cambridge University.[53] Alternatively, the crystal 'seeds' are supposed to have been blown around in the atmosphere as microscopic particles of dust, before settling in crystallising dishes, catalysing the crystallisation of the new substance. An American chemist, C. P. Saylor, commented, that it was as if 'the seeds of crystallisation, as dust, [were] carried upon the winds from end to end of the earth'.[54]

Thus the formation of new kinds of crystals provides one way of testing the hypothesis of morphic resonance. According to the conventional assumption, crystals should *not* form more readily in a laboratory in Australia after they have been made in a laboratory in Britain if visitors from the British laboratory are rigorously excluded, and if dust particles are filtered out of the atmosphere. If they do form more rapidly, this result would favour morphic resonance. I discuss further tests with crystals in my book *A New Science of Life*.[55]

Habit and creativity

Habits alone cannot explain evolution. They are by their very nature conservative. They account for repetition, but not for creativity. Evolution must involve a combination of these two processes: through creativity, new patterns of organisation arise; those that survive and are repeated become increasingly habitual. Some new patterns are favoured by natural selection and others are not.

Creativity is a mystery precisely because it involves the appearance of patterns that have never existed before. Our usual way of explaining things is in terms of pre-existing causes: the cause somehow contains the effect; the effect follows from the cause. If we apply this way of thinking to the creation of a new form of life, a new work of art, or a new idea, we infer that the new pattern of organisation was already present: it was a latent possibility. Given the appropriate circumstances, this latent pattern becomes actual. It is *discovered* rather than created. Creativity consists in the manifestation of eternally pre-existing possibilities. In other words, the new pattern has not been created at all; it has only been manifested in the physical world, whereas previously it was unmanifest.

This in essence is the Platonic theory of creativity. All possible forms have always existed as timeless Forms, or as mathematical potentialities implicit in the eternal laws of nature: 'The possible would have been there from all time, a phantom awaiting its hour; it would therefore have become reality by the addition of something, by some transfusion of blood or life,' as Henri Bergson expressed it.[56] Bergson (1859–1941) was an evolutionary philosopher far ahead of his time, and influenced William James and Alfred North Whitehead. In his most famous book, *Creative Evolution*, he was very clear on what a profound break the concept of evolution made with the habits of Platonic thought:

> The ancients, Platonists to a greater or lesser degree . . . imagined
> that Being was given once and for all, complete and perfect, in
> the immutable system of Ideas; the world which unfolds before

our eyes could therefore add nothing to it; it was, on the contrary, only diminution or degradation; its successive states measured as it were the increasing or decreasing distance between what is, a shadow projected in time, and what ought to be, Idea set in eternity. The moderns, it is true, take a quite different point of view. They no longer treat Time as an intruder, a disturber of eternity; but they would very much like to reduce it to a simple appearance. The temporal is, then, only the confused form of the rational . . . The real becomes once more the eternal, with this simple difference, that it is the eternity of the Laws in which the phenomena are resolved instead of being the eternity of the Ideas which serve them as models.[57]

Eternal Forms or laws seemed appropriate enough in an eternal universe. They are thrown into question by evolution, a process of creative development. Creativity is real; new patterns of organisation appear as the world develops. Everything new that happens is possible in the tautological sense that only the possible can happen. Bergson argued that we need not attribute to these possibilities, which are unknowable until they actually happen, a pre-existent reality transcending time and space.

By contrast, the theory of evolution by natural selection was not Platonic. It was based on observations of fossils and actual living organisms. For Charles Darwin, the source of evolutionary creativity was not beyond nature, in the eternal designs and plans of a machine-making God, the God of Paley's natural theology (see Chapter 1). The evolution of life took place spontaneously. Nature itself, or herself, has given rise to all the myriad forms of life.

Henri Bergson attributed this creativity to the *élan vital* or vital impetus. Like Darwinians, Marxists and other believers in emergent evolution, he denied that the evolutionary process was designed and planned in advance in the mind of a Platonic God. Instead, evolution is spontaneous and creative:

Nature is more and better than a plan in course of realisation. A plan is a term assigned to a labour: it closes the future whose form it indicates. Before the evolution of life, on the contrary, the portals of the future remain wide open. It is a creation that goes on for ever in virtue of an initial movement. This movement constitutes the unity of the organised world – a prolific unity, of an infinite richness, superior to any that the intellect could dream of, for the intellect is only one of its aspects or products.[58]

What difference does it make?

Abandoning the dogma of fixed laws liberates our understanding of evolution. The Big Bang theory locates cosmic creativity at the beginning. In the original miracle, all the laws of nature and all the matter and energy in the universe suddenly arose from nothing, or from the wreckage of a previous universe. By contrast, a radically evolutionary view of nature implies an ongoing creativity, establishing new habits and regularities as nature evolves. Human creativity is part of a vast creative process that has been unfolding through the whole of evolution.

The inheritance of habits by morphic resonance makes a big difference in the understanding of the inheritance of form, learning and memory, as discussed in Chapters 6 and 7.

When chemists make new chemical compounds that, as far as we know, have never existed on earth, they should show an increasing ease of crystallisation as time goes on, as discussed above. But what if these crystals have existed on other planets? If morphic resonance does not fall off with distance, then these new crystals should be influenced by morphic resonance from crystals of the same kind on other planets, and should crystallise readily, without an apparent learning effect.

Hence it should be possible to discover which new chemicals are unique to the earth and which have existed elsewhere. If the rate of crystallisation of, say, a thousand new chemicals is measured systematically, and if, say, 800 show increasing rates of crystallisation while the other 200 do not, we could infer that the latter have existed elsewhere in the universe, but the former have not. Inexpensively, we could find what is truly new on earth, and deduce something about events on other planets, even though we do not know where those planets are.

Questions for materialists

If the laws of nature existed before the Big Bang, and governed the Big Bang from its first instant, where were they?

If the laws and constants of nature all came into being at the moment of the Big Bang, how does the universe remember them? Where are they 'imprinted'?

How do you know that the laws of nature are fixed and not evolutionary?

What is wrong with the idea that nature has habits rather than laws?

SUMMARY

The idea that the 'laws of nature' are fixed while the universe evolves is an assumption left over from pre-evolutionary cosmology. The laws may themselves evolve or, rather, be more like habits. Also, the 'fundamental constants' may be variable, and their values may not have been fixed at the instant of the Big Bang. They still seem to be varying today. There may be an inherent memory in nature. All organisms may participate in a collective memory of their kind. Crystals may crystallise the way they do because they formed that way before; the more crystals of a particular chemical arise in one place, the easier should they crystallise everywhere else on earth, and maybe throughout the universe. Evolution may be the result of an interplay between habits and creativity. New forms and patterns of organisation appear spontaneously, and are subject to natural selection. Those that survive are more likely to appear again as new habits build up, and through repetition they become increasingly habitual.

4

Is Matter Unconscious?

The central doctrine of materialism is that matter is the only reality. Therefore consciousness ought not to exist. Materialism's biggest problem is that consciousness does exist. You are conscious now. The main opposing theory, dualism, accepts the reality of consciousness, but has no convincing explanation for its interaction with the body and the brain. Dualist-materialist arguments have gone on for centuries. In this chapter I suggest how we can move forwards from this sterile opposition.

Scientific materialism arose historically as a rejection of mechanistic dualism, which *defined* matter as unconscious and souls as immaterial, as I discuss below. One important motive for this rejection was the elimination of souls and God. In short, materialists treated subjective experience as irrelevant; dualists accepted the reality of experience but were unable to explain how minds affect brains.

The materialist philosopher Daniel Dennett wrote a book called *Consciousness Explained* (1991) in which he tried to explain away consciousness by arguing that subjective experience is illusory. He was forced to this conclusion because he rejected dualism as a matter of principle:

> I adopt the apparently dogmatic rule that dualism is to be avoided *at all costs*. It is not that I think I can give a knock-down proof that dualism, in all its forms, is false or incoherent, but that, given the way that dualism wallows in mystery, *accepting dualism is giving up* [his italics].[1]

The dogmatism of Dennett's rule is not merely apparent: the rule *is* dogmatic. By 'giving up' and 'wallowing in mystery', I suppose he means giving up science and reason and relapsing into religion and superstition. Materialism 'at all costs' demands the denial of the reality of our own minds and personal experiences – including those of Daniel Dennett himself, although by putting forward arguments he hopes will be persuasive, he seems to make an exception for himself and for those who read his book.

Francis Crick devoted decades of his life to trying to explain consciousness mechanistically. He frankly admitted that the materialist theory was an 'astonishing hypothesis' that flew in the face of common sense: "'You", your joys and your sorrows, your memories and your ambitions, your sense of personal identity and free will, are in fact no more than the behaviour of a vast assembly of nerve cells and their associated molecules.'[2] Presumably Crick included himself in this description, although he must have felt that there was more to his argument than the automatic activity of nerve cells.

One of the motives of materialists is to support an anti-religious worldview. Francis Crick was a militant atheist, as is Daniel Dennett. On the other hand, one of the traditional motives of dualists is to support the possibility of the soul's survival. If the human soul is immaterial, it may exist after bodily death.

Scientific orthodoxy has not always been materialist. The founders of mechanistic science in the seventeenth century were dualistic Christians. They downgraded matter, making it totally inanimate and mechanical, and at the same time upgraded human minds, making them completely different from unconscious matter. By creating an unbridgeable gulf between the two, they thought they were strengthening the argument for the human soul and its immortality, as well as increasing the separation between humans and other animals.

This mechanistic dualism is often called Cartesian dualism, after Descartes (Des Cartes). It saw the human mind as essentially immaterial and disembodied, and bodies as machines made of unconscious matter.[3] In practice, most people take a dualist view for granted, as long as they are not called upon to defend it. Almost everyone assumes that we have some degree of free will, and are responsible for our actions. Our educational and legal systems are based on this belief. And we experience ourselves as conscious beings, with some degree of free choice. Even to discuss consciousness presupposes that we are conscious ourselves. Nevertheless, since the 1920s, most leading scientists and philosophers in the English-speaking world have been materialists, in spite of all the problems this doctrine creates.

The strongest argument in favour of materialism is the failure of dualism to explain how immaterial minds work and how they interact with brains. The strongest argument in favour of dualism is the implausibility and self-contradictory nature of materialism.

The dualist-materialist dialectic has lasted for centuries. The soul-body or mind-brain problem has refused to go away. But before we can move forward, we need to understand in more detail what materialists claim, since their belief-system dominates institutional science and medicine, and everyone is influenced by it.

Minds that deny their own reality

Most neuroscientists do not spend much time thinking about the logical problems that materialist beliefs entail. They just get on with the job of trying to understand how brains work, in the faith that more hard facts will eventually provide answers. They leave professional philosophers to defend the materialist or physicalist faith.

Physicalism means much the same as materialism, but rather than asserting that all reality is material, it asserts that it is physical, explicable in terms of physics, and therefore including energy and fields as well as matter. In practice, this is what materialists believe too. In the following discussion I use the more familiar word materialism to mean 'materialism or physicalism'.

Among materialist philosophers there are several schools of thought. The most extreme position is called 'eliminative materialism'. The philosopher Paul Churchland, for example, claims that there is nothing more to the mind than what occurs in the brain. Those who believe in the existence of thoughts, beliefs, desires, motives and other mental states are victims of 'folk psychology', an unscientific attitude that will in due course be replaced by explanations in terms of the activities of nerves. Folk psychology is a kind of superstition, like belief in demons, and it will be left behind by the onward march of scientific understanding. Consciousness is just an 'aspect' of the activity of the brain. Thoughts or sensations are just another way of talking about activity in particular regions of the cerebral cortex; they are the same things talked about in different ways.

Other materialists are 'epiphenomenalists': they accept rather than deny the existence of consciousness, but see it as a functionless by-product of the activity of the brain, an 'epiphenomenon', like a shadow. T. H. Huxley was an early advocate of this point of view, and in 1874 he famously compared consciousness to 'the steam whistle that accompanies the work of a locomotive engine . . . without influence upon its machinery'.[4] He concluded, 'We are conscious automata.'[5] People might just as well be zombies, with no subjective experience, because all their behaviour is a result of brain activity alone. Conscious experience does nothing, and makes no difference to the physical world.

A recent form of materialism is 'cognitive psychology', which dominated academic psychology in the English-speaking world in the late twentieth century. It treats the brain as a computer and mental activity as information processing. Subjective experiences, like seeing green, or feeling pain, or enjoying music, are computational processes inside the brain, which are themselves unconscious.

Some philosophers, like John Searle, think that minds can emerge from matter by analogy with the way that physical properties can emerge at different levels of complexity, like the wetness of water emerging from the interactions of large numbers of water molecules. In nature, there are indeed many different levels of organisation (Figure 1.1), each of which has new properties that are not present in their parts alone. Atoms have properties over and above nuclear particles and electrons. Molecules have properties over and above atoms: the molecules of water, H_2O, are fundamentally different from uncombined hydrogen and oxygen atoms. Then the wetness of liquid water is not explained by water molecules in isolation, but through their organisation together in liquid water. New physical properties 'emerge' at every level. In the same way, consciousness is an emergent physical property of brains. It is different from other physical processes, but it is physical none the less. Many non-materialists would agree with Searle that consciousness is in some sense 'emergent' but would argue that while mind and conscious agency originate in physical nature, they are qualitatively different from purely material or physical being.

Finally, some materialists hope that evolution can provide an answer. They propose that consciousness emerged as a result of natural selection through mindless processes from unconscious matter. Because minds evolved, they must have been favoured by natural selection and hence they must actually do something: they must make a difference. Many non-materialists would agree. But materialists want to have it both ways: emergent consciousness must do something if it has evolved as an evolutionary adaptation favoured by natural selection; but it cannot do anything if it is just an epiphenomenon of brain activity, or another way of talking about brain mechanisms. In 2011, the psychologist Nicholas Humphrey tried to overcome this problem by suggesting that consciousness evolved because it helps humans survive and reproduce by making us feel 'special and transcendent'. But as a materialist, Humphrey does not agree that our minds have any agency; that is to say, they cannot affect our actions. Instead our consciousness is illusory: he describes it as 'a magical mystery show that we stage for ourselves inside our own heads'.[6] But to say that consciousness is an illusion does not *explain* consciousness: it presupposes it. Illusion is a mode of consciousness.

If all these theories sound unconvincing, that is because they are. They do not even convince other materialists, which is why there are so many rival theories. Searle has described the debate over the last fifty years as follows:

> A philosopher advances a materialist theory of the mind . . . He
> then encounters difficulties . . . Criticisms of the materialist
> theory usually take a more or less technical form, but, in fact,
> underlying the technical objections is a much deeper objection:
> the theory in question has left out some essential feature of the
> mind . . . And this leads to ever more frenzied attempts to stick
> with the materialist thesis.[7]

The philosopher Galen Strawson, himself a materialist, is amazed
by the willingness of so many of his fellow philosophers to deny
the reality of their own experience:

> I think we should feel very sober, and a little afraid, at the power
> of human credulity, the capacity of human minds to be gripped
> by theory, by faith. For this particular denial is the strangest thing
> that has ever happened in the whole history of human thought,
> not just the whole history of philosophy.[8]

Francis Crick admitted that the 'astonishing hypothesis' was not
proved. He conceded that a dualist view might become more plausible. But, he added,

> There is always a third possibility: that the facts support a new,
> alternative way of looking at the mind-brain problem that is
> significantly different from the rather crude materialistic view
> that many neuroscientists hold today and also from the religious
> point of view. Only time, and much further scientific work, will
> enable us to decide.[9]

There is indeed a third way.

Mental matter

Galen Strawson shares the frustration of many contemporary
philosophers with the seemingly intractable problems of materialism and dualism. He has come to the conclusion that there is

only one way out. He argues that a consistent materialism must imply panpsychism, namely the idea that even atoms and molecules have a primitive kind of mentality or experience. (The Greek word *pan* means everywhere, and *psyche* means soul or mind.) Panpsychism does not mean that atoms are conscious in the sense that we are, but only that some aspects of mentality or experience are present in the simplest physical systems. More complex forms of mind or experience emerge in more complex systems.[10]

In 2006, the *Journal of Consciousness Studies* published a special issue entitled 'Does materialism entail panpsychism?' with a target article by Strawson, and responses by seventeen other philosophers and scientists. Some of them rejected his suggestion in favour of more conventional kinds of materialism, but all admitted that their favoured kind of materialism was problematic.

Strawson made only a generalised, abstract case for panpsychism, with disappointingly few details as to how an electron or an atom could be said to have experiences. But, like many other panpsychists, he made an important distinction between aggregates of matter, like tables and rocks, and self-organising systems like atoms, cells and animals. He did not suggest that tables and rocks have any unified experience, though the atoms within them may have.[11] The reason for this distinction is that man-made objects, like chairs or cars, do not organise themselves, and do not have their own goals or purposes. They are designed by people and put together in factories. Likewise rocks are made up of atoms and crystals that are self-organising, but external forces shape the rock as a whole: for example, it may have been split from a larger rock as a boulder rolled down a mountain.

By contrast, in *self-organising* systems, complex forms of experience emerge spontaneously. These systems are at the same time physical (non-experiential) and experiential; in other words, they have experiences. As Strawson put it, 'Once upon a time there was relatively unorganised matter with both experiential and non-experiential fundamental features. It organised into increasingly complex forms, both experiential and non-experiential, by many processes including evolution by natural selection.'[12] Unlike Searle's attempt to explain consciousness by saying that it emerges from totally unconscious, insentient matter, Strawson's proposal is that more complex forms of experience emerge from less complex ones. There is a difference of degree, but not of kind.

Panpsychism is not a new idea. Most people used to believe in it, and many still do. All over the world, traditional people saw the world around them as alive and in some sense conscious or aware: the planets, stars, the earth, plants and animals all had spirits or souls. Ancient Greek philosophy grew up in this context, although some of the earliest philosophers were hylozoists, rather than panpsychists; that is, they saw all things as in some degree alive, without necessarily supposing that they had sensations or experiences. In medieval Europe, philosophers and theologians took for granted that the world was full of animate beings; plants and animals had souls, and stars and planets were governed by intelligences. Today this attitude is usually rejected as 'naïve', or 'primitive', or 'superstitious'. Searle described it as 'absurd'.[13] Yet some of the greatest Western philosophers have supported a panpsychist point of view, for the same reasons as Strawson. Soon after Descartes' philosophy was published, thinkers who opposed his stark dualism were looking for new ways of understanding how minds and bodies were related in all nature, not just in human brains.

Physics and experience

The philosopher Baruch Spinoza (1632–77) argued that everything in nature has both a body and a mind. The mind and body were two aspects of the same underlying reality, which he called *Deus sive natura*, God or Nature, and they changed in parallel. In general, the greater the complexity of a body's interaction with the world, the greater the complexity of the corresponding mind. The most basic aspect of substances at all levels of complexity was what Spinoza called *conatus*, a Latin word meaning 'striving' that was both physical and mental. In his own words:

Each thing, as far as it can by its own power, strives to persevere
in its being . . . The striving by which each thing strives to persevere in its own being is nothing but the actual essence of the
thing.[14]

This striving was equivalent to appetite, and desire was appetite,
with consciousness of it. Spinoza proposed that the transition to a
state of greater power or perfection in any individual was experienced as pleasure, and a diminution of power as pain.[15]

Gottfried Leibniz (1646–1716) was a polymath and mathematician who invented the infinitesimal calculus independently of
Isaac Newton. Both Newton and Leibniz had a vision of holistic
interconnectedness. But while Newton thought that matter was
made up of unconscious particles that attracted every other particle in the universe through gravitational attraction, Leibniz
proposed that the ultimate elements of the universe were interre-
lated through consciousness. He called these ultimate units
monads, which were both physical centres of force and mental
centres of experience, each reflecting the universe. As Leibniz
put it, 'Each monad is a living mirror . . . which represents the
universe from its own point of view and is as ordered as the
universe itself.'[16] The monads had two primary qualities, 'perception' and 'appetite'. Perceptions were the changing internal states
of the monads, which arose from their appetites, which in turn
arose from their need to reflect the universe.[17] Monads were unities
of force and mind, whereas Newton's particles were merely unconscious centres of force.

In the eighteenth century some of the leading proponents of
Enlightenment materialism combined a mechanistic theory of
life, with the belief that matter itself had sensations and feelings.
Julien de la Mettrie, the author of a famous book called *L'Homme
Machine* (Man, a machine, 1748), denied the existence of the soul,
but he animated the matter of the body instead, endowing it with
feeling.[18]

Denis Diderot, a prominent Enlightenment philosopher,
extended the realm of subjectivity to all matter, not just living

organisms. In 1769 he wrote, 'The faculty of sensation . . . is a general and essential quality of matter.'[19] He spoke of 'intelligent particles', and added, 'From the elephant to the flea, from the flea to the sensitive living atom, the origin of all, there is no point in nature but suffers and enjoys.'[20]

From around 1780 to 1880, panpsychism was especially influential in Germany. The philosopher Johann Herder (1744–1803) argued that force or energy was the underlying principle of reality, manifesting both mental and physical properties. The poet Wolfgang von Goethe, a friend of Herder's, envisaged two great driving forces in nature: polarity and intensification. Polarity was associated with the material dimension, as 'a state of constant attraction and repulsion', and intensification gave a spiritual dimension, which was one of 'a state of ever-striving ascent', a kind of evolutionary imperative. On the principle that there could be no matter without mind and no mind without matter, 'matter is also capable of undergoing intensification, and spirit cannot be denied its attraction and repulsion'.[21]

The philosopher Arthur Schopenhauer argued in *The World as Will and Idea* (1819) that all things possess a will, expressed through desires, feelings and emotions. Material bodies were 'objectifications' of will. Physical forces, including gravitation, magnetic attraction and repulsion, were manifestations of the will in nature.

Many other nineteenth-century philosophers in the German-speaking world advocated similar views, but two are especially important. The Austrian philosopher of science Ernst Mach (1838–1916), who influenced Albert Einstein's theory of relativity, explicitly rejected a mechanistic conception of matter, and wrote, 'Properly speaking, the world is not composed of "things" . . . but of colours, tones, pressures, spaces, times, in short what we ordinarily call individual sensations.'[22] And Ernst Haeckel, Germany's most prominent advocate of Darwin's theory of evolution, wrote in 1892, 'I regard *all* matter as *ensouled*, that is to say as endowed with *feeling* (pleasure and pain) and motion.' He claimed that all living creatures, including microbes, possess 'conscious psychic action'. Inorganic matter also had a mental aspect, but 'I conceive the elementary psychic qualities of sensation and will, which may be attributed to atoms, to be *unconscious*.'[23]

In the United States, the pioneering psychologist William James advocated a form of panpsychism in which individual minds and a hierarchy of lower- and higher-order minds constituted the reality of the cosmos.[24] The philosopher Charles Sanders Peirce saw the physical and mental as different aspects of underlying reality: 'All mind more or less partakes of the nature of matter . . . Viewing a thing from the outside . . . it appears as matter. Viewing it from the inside . . . it appears as consciousness.'[25]

In France, the philosopher Henri Bergson took this tradition of thought to a new level by emphasising the importance of memory. All physical events contain a memory of the past, which is what enables them to endure. The unconscious matter of mechanistic physics was assumed by Bergson's contemporaries to persist unchanged until acted on by external forces; matter lived in an eternal instant, and had no time within it. Bergson argued that mechanistic physics treated changes cinematographically, as if there were a series of static, frozen moments, but for him, this kind of physics was an abstraction that left out the essential feature of living nature: 'Duration is essentially a continuation of what no longer exists into what does exist. This is real time, perceived and lived . . . Duration therefore implies consciousness; and we place consciousness at the heart of things for the very reason that we credit them with a time that endures.'[26]

Even some of the most influential modern materialists cannot resist endowing biochemical systems with subjectivity. Richard Dawkins's 'selfish genes' are an example of animated matter. But whereas Dawkins's molecular vitalism is avowedly a rhetorical device, his philosophical colleague, Daniel Dennett, tried to conjure a primitive kind of consciousness out of genes or replicators by endowing them with an 'interest' in self-replication: 'When an entity arrives on the scene capable of behaviour that staves off, however primitively, its own dissolution and decomposition, it brings with it into the world its "good". That is to say, it creates a point of view.'[27]

Occasions of experience

The leading panpsychist philosopher in the English-speaking world was Alfred North Whitehead, who started his career as a mathematician at Trinity College, Cambridge, where he taught Bertrand Russell. Together they co-authored *Principia Mathematica* (1910–1913), one of the most important works in twentieth-century mathematical philosophy. Whitehead then developed a theory of relativity that made almost identical predictions to Einstein's, and both theories were confirmed by the same experiments.

Whitehead was probably the first philosopher to recognise the radical implications of quantum physics. He realised that the wave theory of matter destroyed the old idea of material bodies as essentially spatial, existing at points in time, but without any time within them. According to quantum physics, every primordial element of matter is 'an organised system of vibratory streaming of energy'.[28] A wave does not exist in an instant, it takes time; its waves connect the past and the future. He thought of the physical world as made up not of material objects but *actual entities* or *events*. An event is a happening or a becoming. It has time within it. It is a process, not a thing. As Whitehead put it, 'An event in realising itself displays a pattern.' The pattern 'requires a duration involving a definite lapse of time, and not merely an instantaneous moment'.[29]

As Whitehead made clear, physics itself was pointing to the conclusion that Bergson had already reached. There is no such thing as timeless matter. All physical objects are processes that have time within them, an inner duration. Quantum physics shows that there is a minimum time period for events, because everything is vibratory, and no vibration can be instantaneous. The fundamental units of nature, including photons and electrons, are temporal as well as spatial. There is no 'nature at an instant'.[30]

Perhaps the most astonishing and original feature of Whitehead's theory was his new perspective on the relationship between mind and body as a relationship in *time*. The usual way of conceiving this relationship is spatial: your mind is inside your body, while the physical world is outside. Your mind sees things from within; it has an inner life. Even from the materialist point of view, the mind is literally 'inside' – inside the brain, insulated within the darkness of the skull. The rest of the body and the entire external world are 'outside'.

By contrast, for Whitehead mind and matter are related as phases in a process. Time, not space, is the key to their relationship. Reality consists of moments in process, and one moment informs the next. The distinction between moments requires the experiencer to feel the difference between the moment of now and past or future moments. Every actuality is a moment of experience. As it expires and becomes a past moment, it is succeeded by a new moment of 'now', a new subject of experience. Meanwhile the moment that has just expired becomes a past object for the new subject – and an object for other subjects too. Whitehead summed this up in the phrase, 'Now subject, then object.'[31] Experience is always 'now', and matter is always 'ago'. The link from the past to the present is physical causality, as in ordinary physics, and from the present to the past is feeling or, to use Whitehead's technical term, 'prehension', meaning, literally, seizing or grasping.

According to Whitehead, every actual occasion is therefore both determined by physical causes from the past, and by the self-creative, self-renewing subject that both chooses its own past and chooses among its potential futures. Through its prehensions it selects what aspects of the past it brings into its own physical being in the present, and also chooses among the possibilities that determine its future. It is connected to its past by selective memories, and connected to its potential future through its choices. Even the smallest possible processes, like quantum events, are both physical and mental; they are oriented in time. The direction of physical causation is from the past to the present, but the direction of mental activity runs the other way, from the present into the past through prehensions, and from potential futures into the present. There is thus a time-polarity between the mental and physical poles of an event: physical causation from past to present, and mental causation from present to past.

Whitehead was not proposing that atoms are conscious in the same way that we are, but that they have experiences and feelings. Feelings, emotions and experiences are more fundamental than human consciousness, and every mental event is informed and causally conditioned by material events, which are themselves composed of expired experiences. Knowing can happen only because the past streams into the present, forming it and shaping it, and at the same time the subject chooses among the possibilities that help determine its future.[32]

Whitehead's philosophy is notoriously difficult to follow, especially in his key book *Process and Reality* (1929), but his insights about the temporal relationship of mind and matter point to a way forward and are well worth trying to understand, even if they are very abstract. One of his modern exponents, Christian de Quincey, has described his idea as follows:

> Think of reality as made up of countless gazillions of 'bubble moments', where each bubble is both physical and mental – a bubble or quantum of *sentient energy* . . . Each bubble exists for a moment and then *pops!* and the resulting 'spray' is the objective 'stuff' that composes the physical pole of the next momentary bubble . . . Time is our experience of the ongoing succession of these momentary bubbles of being (or bubbles of *becoming*) popping in and out of the present moment of *now*. We feel this succession of moments as the flow of the present slipping into the past, always replenished by new moments of 'now' from an apparently inexhaustible source we objectify as the future . . . The future does not exist except as *potentials* or possibilities in the present moment – in experience – which is always conditioned by the objective pressure of the past (the physical world). Subjectivity (consciousness, awareness) is what-it-feels-like to experience these possibilities, and choosing from them to create the next new moment of experience.[33]

The relation of conscious experience to time has been investigated experimentally with intriguing results.

Conscious experience and brain activity

Many philosophers have speculated about the relationship between the mind and the brain, but the neuroscientist Benjamin Libet and his team in San Francisco investigated it experimentally by measuring changes in the brain and the timing of conscious experiences.

First, Libet's group stimulated their human subjects either by flashes of light or by a rapid sequence of mild electric pulses applied to the back of the hand. If the stimulus was short, less than about half a second (500 milliseconds), the subjects were unconscious of it, even though the sensory cortex of their brains responded. But if the stimulus went on for more than 500 milliseconds, the subjects became consciously aware of it. So far, so good. The need for a minimum duration of stimulus is not in itself surprising. What *is* surprising is that the subjects' conscious awareness of the stimulus began not after 500 milliseconds but when the stimulus started. In other words, it took half a second for the stimulus to be experienced subjectively, but this subjective experience moved backwards to when the stimulus was first applied. 'There is an automatic subjective referral of the conscious experience backwards in time . . . The sensory experience would be "antedated" from the actual delayed time at which the neuronal state becomes adequate to elicit it; and the experience would appear subjectively to occur with no significant delay.'[34]

Second, Libet investigated what happened when people made free conscious choices. He measured the electrical activity of their brains by means of an electroencephalograph (EEG), with small electrodes placed on the surface of the head. The subjects sat quietly, and were asked to flex one of their fingers or push a button whenever they felt like doing so. They also noted when they decided or felt the wish to do so. This conscious decision occurred about 200 milliseconds before the finger movement. This seems straightforward – the choice preceded the action. What was remarkable was that electrical changes began in the brain about

300 milliseconds *before* any conscious decision was made.[35] These changes were called the 'readiness potential'.

For some neuroscientists and philosophers, Libet's finding seemed like the ultimate experimental proof that free will is an illusion. The brain changed first, and about a third of a second later, conscious awareness followed the choice, rather than initiating it. Therefore unconscious physical processes caused the 'decision', not free will.[36]

Libet himself took a different view. He suggested that in the time between conscious awareness of the desire to act and the actual movement – a gap of 200 milliseconds – there was an opportunity for the person's conscious mind to veto the decision. Instead of free will, we have 'free won't'. This conscious decision depended on what Libet called a 'conscious mental field' (CMF), which emerged from brain activities but was not itself physically determined by them. The CMF acted on the activities of the brain, perhaps by influencing otherwise random or indeterminate events in the nerve cells. This field also helped integrate the activities of different parts of the brain and had the property of 'referring back' subjective experiences, and thus worked backwards in time.[37]

> The CMF would unify the experience generated by the many neural units. It would also be able to affect certain neural activities and form a basis for conscious will. The CMF would be a new 'natural' field. It would be a non-physical field, in the sense that it could not be directly observed or measured by any external physical means. That attribute is, of course, the well-known feature of conscious subjective experience, which is only accessible to the individual having that experience.[38]

To go one step further than Libet did, if the mental field worked backwards in time on the activity of the nerves, then the conscious mental field could *cause* the readiness potential that preceded it. Mental causation would work from the future towards the past, while physical causation works from the past towards the future.

The materialist interpretation of Libet's finding assumes that causation works in only one direction, from the past towards the future. But if mental causation works in the opposite direction, then the conscious choice could trigger the readiness potential. In Chapter 9, I discuss further experimental evidence for a backwards-in-time flow of influences from future mental states.

Conscious and unconscious minds

There are at least two senses of the word 'unconscious'. One means totally devoid of mind, experience and feeling, and this is what materialists mean when they say matter is unconscious. Physicists and chemists treat the systems they study as unconscious in this absolute sense. But a very different meaning of 'unconscious' is implied by the phrase 'unconscious mind'. Most of our own mental processes are unconscious, including most of our habits. When driving a car we can carry on a conversation while our perceptions of the road and other vehicles affect our responses, without our being consciously aware of all our movements and choices. When I come to a familiar road junction, I may turn right automatically, because this is my habitual route. I am choosing among possibilities, but choosing on the basis of habit. By contrast, if I am driving in an unfamiliar town and trying to find my way with the help of a map, my choice when I come to a junction depends on conscious deliberation. But only a small minority of our choices are conscious. Most of our behaviour is habitual, and habits by their very nature work unconsciously.

Like humans, animals are largely creatures of habit. Yet the fact that they are not conscious of most of their actions – as we are not conscious of most of our own – does not mean they are mindless machines. They have a mental aspect as well as a physical aspect, and their mental aspect is shaped by their habits, feelings and potentialities, among which they choose, unconsciously or consciously.

It may not make much sense to suggest that electrons, atoms and molecules make conscious choices, but they may make unconscious choices on the basis of habits, just as we do and animals do.

According to quantum theory, even elementary particles like electrons have many alternative future possibilities. The calculation of their behaviour by physicists involves taking all their possible futures into account.[39] Electrons are physical in that they re-enact elements of their past; but they also have a mental pole in that they relate this re-enactment of the past to their future potentialities, which in some sense work backwards in time.

But can we meaningfully say that electrons have experiences, feelings and motivations? Can they be attracted towards one possible future, or repelled by another? The answer is 'yes'. For a start, they are electrically charged; they 'feel' the electric field around them; they are attracted towards positively charged bodies, and repelled by those with negative charges. Physicists model their behaviour mathematically without supposing that their feelings, attractions and repulsions are anything other than physical forces, or that their individually unpredictable behaviour is governed by anything other than chance and probability. Materialists would say that only by fanciful metaphors can they be seen to have feelings or experience. But some physicists think differently, like David Bohm and Freeman Dyson. Bohm observed, 'The question is whether matter is rather crude and mechanical or whether it gets more and more subtle and becomes indistinguishable from what people have called mind.'[40] Freeman Dyson wrote,

> I think our consciousness is not just a passive epiphenomenon carried along by the chemical events in our brains, but is an active agent forcing the molecular complexes to make choices between one quantum state and another. In other words, mind is already inherent in every electron, and the processes of human consciousness differ only in degree but not in kind from the processes of choice between quantum states which we call 'chance' when they are made by an electron.[41]

These are difficult questions, and raise all sorts of further questions about the meaning of words like 'feeling', 'experience' and 'attraction'. Are they metaphorical when applied to quantum systems? Perhaps. But we do not have a choice between metaphorical and non-metaphorical thinking. There are no metaphor-free zones in science. The whole of science is suffused with legal metaphors, as in 'laws of nature', materialist theories of mind by computer metaphors, and so on. But the issues are not merely literary or rhetorical, but scientific. As Bergson and Whitehead made clear, and as Libet showed by experiment, the mental and physical aspects of material bodies have different relationships to time and to causation.

I return to a discussion of influences flowing from the future towards the past in Chapter 5 in the context of purposes in nature.

What difference does it make?

The question 'Is Matter Unconscious?' is not just an abstract, intellectual question. It makes a huge difference. It affects the way we relate to other people and to the world and shapes our experience of ourselves. If materialism is true, all bodies, including yours and mine, are essentially unconscious. Your subjective experiences emerge from your brain as epiphenomena, or else they are merely an aspect of the physical activity of your brain, but they cannot have any effects. Your thoughts, desires and decisions cannot interfere with regular physical causality. Your choices are illusory. Materialism promises that, at some time in the future, all human behaviour and beliefs, including the belief in materialism, will be fully explained by the physical and chemical mechanisms of human brains, together with random events inside and outside human bodies.

But what if these materialist beliefs are delusions? Perhaps you are really free to choose your beliefs on the basis of arguments, evidence and experience. Perhaps you are really conscious. Perhaps other animals are conscious too, and capable of free choice to some degree. Maybe all organisms, physical and biological, have experiences and feelings, including atoms, molecules, crystals, cells, tissues, organs, plants, animals, societies of organisms, ecosystems, planets, solar systems and galaxies.

It makes a big difference if you think of yourself as a zombie-like mechanism in an unconscious mechanical world, or as a truly conscious being capable of making choices, living among other beings with sensations, experiences and desires.

Questions for materialists

Do you believe that your own consciousness is merely an aspect or epiphenomenon of the activity of your brain?

If consciousness does nothing, why has it evolved as an evolutionary adaptation?

Do you agree with the materialist philosopher Galen Strawson that materialism implies panpsychism?

Is your own belief in materialism determined by unconscious processes in your brain, rather than reason, evidence and choice?

SUMMARY

In the mechanistic science of the seventeenth century, matter was defined as unconscious, and conscious minds were confined to human beings, along with spirits, angels and God. There was a duality of spirit and matter. No one could satisfactorily explain how non-physical minds could interact with material brains, and materialists rejected the existence of these mysterious immaterial entities, leaving only unconscious matter. But since we ourselves are conscious, this elimination of minds created a big problem for materialists, who have tried to explain human consciousness away or dismiss it as illusory. But instead of assuming that materialism and dualism are the only options, some philosophers have explored the idea that all self-organising material systems have a mental as well as a physical aspect. Their minds relate them to their future goals and are shaped by memories of their past, both individual and collective. The relationship of minds to bodies is more to do with *time* than space. Minds choose among possible futures, and mental causation runs in the opposite direction from energetic causation, from virtual futures towards the past, rather than from the past towards the future.

5

Is Nature Purposeless?

Purposes relate to ends or goals or intentions, conscious or unconscious. They link organisms to their potential futures. The word 'purpose' comes from the Latin *proponere*, meaning to propose or put forward. The word 'intend' comes from the Latin *intendere*, to stretch into. The word 'goal' comes from the Middle English *gol*, a boundary or limit. The Greek word for 'end', *telos*, is the root of 'teleology', the study of ends or goals.

These words all point towards a difficult-to-understand concept. Purposes exist in a virtual realm, rather than a physical reality. They connect organisms to ends or goals that have not yet happened; they are *attractors*, in the language of dynamics, a branch of modern mathematics. Purposes or attractors cannot be weighed; they are not material. Yet they influence material bodies and have physical effects. Your activities as you pursue your goals are objective phenomena that can be filmed and measured. A male dog straining at the leash towards a bitch on heat exerts a force, quantifiable by incorporating a spring balance into the leash. The dog's desire has a measurable force and direction. Purposes or motives are causes, but they work by pulling towards a virtual future rather than pushing from an actual past.

In mainstream medieval philosophy, following Aristotle and Thomas Aquinas, all living organisms were thought to have their own ends or purposes, given by their souls. The fundamental purposes of animals and plants were to grow, maintain themselves and reproduce. Their ends or goals were called 'final causes' and worked by attraction. The *telos*, the goal, of an oak seedling was to be an oak tree, reproducing itself. Final causes pulled from the future by attraction, while moving causes worked from the past by pushing.

The mechanistic revolution in seventeenth-century science abolished ends, purposes, goals and final causes. Everything was to be explained mechanically, by matter being pushed from the past, as in billiard-ball physics, or by forces acting in the present, as in gravitation. This four-hundred-year-old doctrine is still an article of faith in the creed of science, but it does not fit the facts. Therefore scientists keep reinventing ends or goals in disguised forms.

The purposes of living organisms

Machines, unlike living organisms, do not have their own internal purposes. Unlike a horse, a car has no desire of its own to go to one place rather than another. A computer has no purposes of its own, but carries out programs designed to serve the purposes of its human user. A guided missile does not choose its own goal; its target is programmed into it, unlike a racing pigeon that spontaneously navigates towards its home. Machines fulfil human purposes, which are external to the machinery, but living organisms, including humans, have their own purposes, ends and goals. As discussed below, the ends are expressed in the first place by their *morphogenesis*, the coming into being of their bodily forms (from the Greek *morphe*, form, and *genesis*, coming-into-being), as in the growth of a beech tree from a seed, or of a kingfisher from an egg.

The mechanistic philosophy abolished final causes, and all nature became devoid of purposes. Biology students learn to explain away purposes in terms of neo-Darwinian evolution: an eye is not for the purpose of seeing but the product of chance genetic mutations and natural selection; eyes evolved because they enabled animals that could see to survive and reproduce better than organisms that could not. The problem with this kind of explanation is that it does not explain the purposiveness of living organisms: it presupposes it. Living organisms exist because their ancestors were already purposive, in that they were able to grow, survive and reproduce. Features that enabled them to do so better were favoured by natural selection, but these fundamental goal-directed activities were already present in the first living cells.

For Descartes and many other scientists, humans still had purposes even though the rest of nature did not. Humans had rational souls above and beyond material nature; they were unique in having conscious minds and purposeful behaviour. Humans were exceptions to the rest of nature. But materialism rejects this doctrine. Humans are not radically different from the rest of nature; there are no such things as immaterial human souls. There are only brains working mechanically.

Nevertheless, people still have purposes, and animals and plants behave in goal-directed ways. So purposes keep coming back, repackaged in words like 'teleonomy', or in the goals of 'selfish genes', which Richard Dawkins imagines are motivated by an irrepressible desire to replicate themselves: 'They are in you and me; they created us, body and mind; and their preservation is the ultimate rationale for our existence.'[1]

Most biologists are split between a practical acceptance of teleology or teleonomy, and a rejection of it in the interests of mechanistic ideology. In most modern biology, the subject is enmeshed in a confusing mixture of teleological rhetoric and pious denial, a muddle that is made worse by confusing two meanings of purpose: first, the purposes of living beings in growing, maintaining themselves and reproducing, passing through their life cycles, usually repeating ancestral, inherited patterns; second, the question of whether the evolutionary process as a whole has any goals or purposes. These are separate questions, and I defer a discussion of possible evolutionary purposes to the end of this chapter.

Living organisms are not unique in having their activities directed towards ends. A falling stone's activity is directed, in the sense that it is attracted towards the ground where it comes to rest. A piece of iron in the presence of a magnet is attracted towards the magnet until it comes as close to it as possible. Gravitational, magnetic and electrical attractions all give rise to limited kinds of directed activity. Living organisms go further.

In his classic book *The Directiveness of Organic Activities* (1945), the biologist E. S. Russell summarised the general features of goal-directed activity in living organisms:

1. When the goal is reached, action ceases: the goal is normally a terminus of action.
2. If the goal is not reached, action usually persists.
3. Such action may be varied, and if the goal cannot be reached in the usual way, it may be reached in another.
4. The same goal can be reached from different beginnings.
5. Goal-directed activity is affected by external conditions, but not determined by them.

One example of the way in which the same goal can be reached from different beginnings is the development of a dragonfly egg after half has been destroyed (Figure 5.1). The posterior part of the egg normally gives rise to the posterior part of the embryo, but if the anterior part of the egg is destroyed, it gives rise to a small but complete embryo. Likewise, in regeneration, a complete organism can be restored from a part: think, for example, of the way that cuttings from a willow tree can each give rise to a new tree. If a flatworm is cut into pieces, each piece can regenerate a new flatworm.

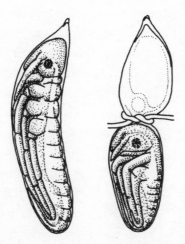

FIGURE 5.1 Left, a normal embryo of the dragonfly *Platycnemis pennipes*. Right, a small but complete embryo formed from the posterior half of an egg tied around the middle soon after laying. (After Weiss, 1939)

Even single cells have astonishing regenerative abilities. *Acetabularia*, the mermaid's wineglass, is a single-celled green alga about five centimetres long, with three main parts: root-like structures called rhizoids that attach it to a rock, a stem, and a cap about a centimetre wide (Figure 5.2). This very large cell has a single nucleus in one of the rhizoids. As the plant grows, its stem lengthens, it forms a series of whorls of hairs that later drop off, and finally forms the cap. If the cap is cut off by snipping the stem in two, after the cut has healed, a new tip grows and the stem forms a series of whorls of hairs and then a new cap, in a similar way to the normal pattern of growth. This can happen over and over again if the cap is cut off repeatedly.[2]

As discussed in the following chapter, the usual assumption is that genes somehow control or 'program' the development of form, as if the nucleus, containing the genes, is a kind of brain

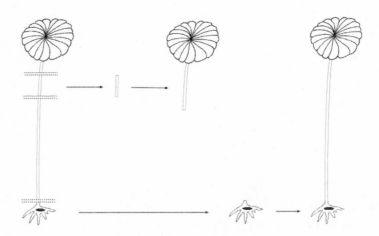

FIGURE 5.2. Regeneration of the alga *Acetabularia mediterranea*, an unusually large single-celled organism, up to 5cm tall, containing a green cap at the top of a long stalk, anchored at the base by root-like rhizoids. There is a large nucleus (shown as a black oval) in the basal part of the cell. When the stalk is cut off near the bottom, the basal part of the cell regenerates a new stalk and cap (shown on the right). When a part of the upper stalk is cut out, it grows a new cap and more stalk, even though it contains no nucleus.

controlling the cell. But *Acetabularia* shows that morphogenesis can take place without genes. If the rhizoid containing the nucleus is cut off, the alga can stay alive for months, and if the cap is cut off, it can regenerate a new one. Even more remarkable, if a piece is cut out of the stem, after the cuts have healed, a new tip grows from the end where the cap used to be and makes a new cap (Figure 5.2).[3] Morphogenesis is goal-directed, and moves towards a morphic attractor even in the absence of genes.

Animal behaviour

Like morphogenesis, animal behaviour is goal-directed, and the instincts of animals can be thought of as pulled towards attractors that help the animals' growth, survival and reproduction, as individuals and as members of social groups, as in a hive of bees. But the fact that animal behaviour is goal-directed does not imply that animals' purposes are conscious, any more than the goal-directed growth of *Acetabularia* means that this alga is conscious.

Instinctive behaviour often consists of chains of more or less stereotyped patterns of behaviour, fixed-action patterns. The end point of one fixed-action pattern may be the starting point of the next. The end points of a chain of fixed-action patterns are called consummatory acts, for example, swallowing food.

As in the development of form, animals have an inherent capacity to adjust or regulate their behaviour so that the end-point is achieved in spite of disturbances. Ethologists, scientists who study animal behaviour, have observed that many fixed-action patterns show a 'fixed' component and an 'orienting' component that is relatively flexible. For example, a greylag goose will retrieve an egg that has rolled out of the nest by putting its bill in front of the egg and rolling it back towards the nest. As the egg is being rolled, its wobbling movements are compensated by appropriate side-to-side movements of the bill.[4] These compensatory movements occur in a flexible way, in response to the movements of the egg, directed towards the fixed goal of rolling the egg back to the nest.

The similarities of goal-directed activity in behaviour and morphogenesis are clearest in the building of nests. For example, female mud
wasps of a *Paralastor* species in Australia build underground nests by
excavating a narrow hole about three inches long and a quarter of an

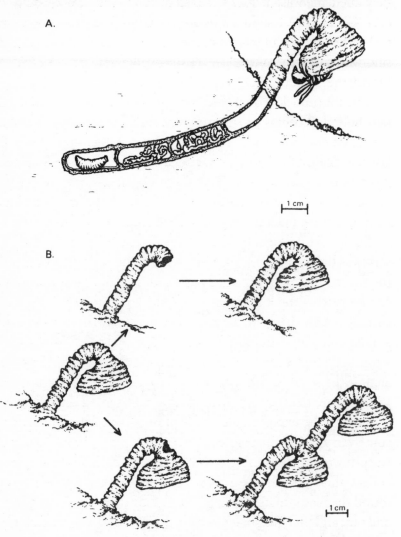

FIGURE 5.3 A: The nest, stocked with food, of the wasp *Paralastor*. B:
The repair of funnels by *Paralastor* wasps. Above, the construction
of a new funnel after the experimenter has removed the old one.
Below, the extra stem and funnel made by the wasp in response to
a hole above the normal funnel. (After Barnett, 1981)

inch wide in a bank of hard, sandy soil. They then line this with mud made from soil near the nest; the wasp releases water from her crop on to the soil, which she then rolls into a ball with her mandibles, carries into the hole and uses to line the walls. After the hole has been fully lined, the wasp begins to construct a large and elaborate funnel over the entrance, building it up from a series of mud pellets (Figure 5.3A). The function of this funnel appears to be the exclusion of parasitic wasps, which cannot get a grip on the smooth inside of the funnel: they fall out when they try to enter.

After the funnel is completed, the wasp lays an egg at the end of the nest hole, and begins provisioning the nest with caterpillars, which are sealed into cells, each about three-quarters of an inch long. The last cell, nearest the entrance, is often sealed off empty, possibly as a protection against parasites. The nest hole is then sealed with a plug of mud, and the wasp destroys the carefully constructed funnel, leaving nothing but a few scattered fragments lying on the ground.

This is a sequence of fixed-action patterns. The end-point of each serves as the stimulus for the next. As in embryonic development, if the normal pathway of activity is disturbed, the same end-points can be reached by a different route. For example, in experiments carried out in the wild, a researcher broke off funnels that were almost complete while the wasps were away collecting mud. The wasps rebuilt them to their original form; the funnels were regenerated. If the scientist broke them off again, the wasps rebuilt them. With one particular wasp this process was repeated seven times.[5]

Second, the experimenter stole almost-complete funnels from some wasps and transplanted them to other nest holes, where the wasps were just beginning to build their funnels and had gone to collect mud. When the wasps returned with their pellets and found the ready-made funnels, they examined them briefly inside and out, and then finished building them as if they were their own.

Third, sand was heaped around funnel stems while they were being constructed. The stems are normally about an inch long. If a nearly completed one was buried until only a fraction of an inch was showing, the wasp went on building up the stem until it was once again about an inch above the ground.

Finally, the researcher made holes in the funnels at different stages of construction. The wasps detected the damage at once, and repaired it with strips of mud.

The most interesting behaviour occurred in response to a type of damage that would probably never happen under natural conditions: the researcher made a circular hole in the neck of the funnel after the bell had been built. The wasps soon noticed these holes and examined them carefully from inside and outside, but they were unable to repair them from the inside because the surface was too slippery. After some delay they started adding mud to the outside of the hole. This is what they do when they start constructing a funnel over the entrance hole of the nest. The holes in the neck of the funnel acted as a sign stimulus for funnel construction, and the wasps built a complete new funnel (Figure 5.2B).

Goal-directedness enables animals to reach their goals in spite of unexpected disturbances, just as developing embryos regulate after damage and produce normal organisms, and just as plants and animals regenerate lost structures.

Attractors

In many models of change, the end or goal is implicitly seen as an attractor by analogy with gravitation. In chemistry, for example, processes of change are modelled in terms of potential wells (Figure 5.4). A system is attracted to the lowest point, its energy minimum.

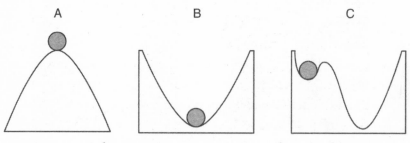

FIGURE 5.4 A diagrammatic representation of an unstable system (A), a stable system in a potential well (B) and a partially stable system (C). The metaphor is gravitational: the ball tends to roll to the lowest position, which has the least potential energy.

In mathematical models of dynamics, goals or ends are represented by *attractors*. Attractors lie in *basins of attraction*. The primary metaphor is a basin into which small balls are thrown. The balls roll around the basin at different speeds and angles but they all end up at the same place, the bottom of the basin, which is the attractor. The plausibility of this metaphor comes from the fact that the basin's bottom is indeed an attractor – a gravitational attractor.

In the mid-twentieth century, the biologist Conrad Waddington described the goal-directed nature of embryonic development in terms of attractors in an 'epigenetic landscape' (Figure 5.5). Each of the end-points represented an organ, such as an eye or a kidney, towards which a part of the embryo developed. The valleys represented the usual pathways of change through which the organ developed. The developmental process was represented by balls rolling along these canalised pathways of change, or *chreodes*, as Waddington called them (from the Greek *chre* = it is necessary and *hodos* = path). One advantage of this model is that it accounts naturally for the development of normal organs even if development is disturbed. If the ball is pushed up the side of a valley, it still rolls towards the attractor when it is released. Waddington thought these epigenetic landscapes represented morphogenetic, or form-shaping, fields.

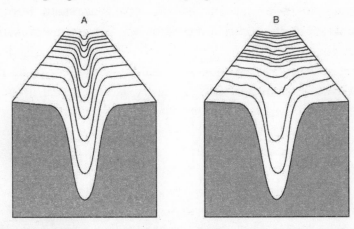

FIGURE 5.5 Diagrammatic representation of a deeply canalized chreode (A) and a chreode that is weakly canalized in the initial stages (B). A ball would roll down the valley towards the end point, which is the attractor.

Once again the attraction in these epigenetic models is analogous to gravitation. Developing systems are attracted towards their ends or goals. They are not only pushed from the past, they are pulled from the future.

In the 1970s and 1980s, the French mathematician René Thom took Waddington's ideas further using dynamical topological models. Whereas Waddington's models were in the form of simple diagrams, Thom's were technical and depended on a branch of mathematics called differential topology, the study of smooth surfaces and their transformations into objects with different spatial properties. His models were also dynamical, in the technical sense of dynamics as the study of change in time, and were situated in multi-dimensional phase spaces. The technical details of Thom's work are hard to understand even for many mathematicians, but he used them to model processes of development in terms of attractors in morphogenetic fields, drawing the developing structures of animals and plants along chreodes towards their developmental goals, such as the structure of an eye or a leaf.[6] The attractors within these fields helped explain the regeneration of lost or damaged structures.

Thom also modelled animal behaviour in terms of attractors. For example, in the capture chreode, a predator seeks, finds and captures food, ending in its ingestion – a consummatory act, in the language of ethology.[7]

Are attractors in morphogenetic fields just abstract mathematics? Or do morphogenetic fields really exert a causal influence, drawing organisms towards their goals? Is there another kind of causation in nature, over and above the forces and fields already known to physics? I think there is, and I think it is related to the flow of influence from the virtual future towards the present discussed in Chapter 4. Causation from virtual ends or attractors working 'backwards' in time fits well with Whitehead's temporal distinction between mind and matter, with mental causes working 'backwards' towards the past. Mental causation flows backwards from the realm of possibilities in the virtual future, and interacts in the present with the energy flowing forward from the past, resulting in observable physical events. The push of energy from the past and the pull from virtual futures overlap in the present, as they do for a ball rolling around a basin.

How can virtual goals exert a causal influence working 'backwards' in time? Is this causal influence confined to the virtual realm to potentialities rather than actualities. Or, in addition can there be a flow of influence from future events to their predecessors?

The possibility of influences flowing backwards from the physical future may not seem worth considering. Most people assume that time-reversed causation is scientifically impossible. But, surprisingly, most of the laws of physics are reversible, and work just as well from the future to the past as from the past to the future. In James Clark Maxwell's classical equations for electromagnetic waves, put forward in 1864, there are two answers that describe the movement of light waves. In one answer, the waves move at the speed of light from the present to the future, as in the conventional understanding of causation. But in the other answer, the waves move from the present into the past at the speed of light, in the opposite direction to ordinary causation. These waves moving backwards in time are called 'advanced waves'. They imply influences working backwards in time. Advanced waves are part of the mathematics of electromagnetism, but physicists ignore them because they are regarded as 'non-physical'.

However, some interpretations of quantum mechanics allow for physical influences working backwards in time or, in other words, causal influences from the future. In Richard Feynman's interpretation, a positron, the anti-particle of an electron, can be thought of as an electron moving 'backwards' in time. And in the 'transactional' interpretation of quantum mechanics,[8] quantum processes are seen as standing waves between emitters and absorbers, with forward-in-time waves moving from emitter to absorber, and backward-in-time waves from absorber to emitter. At the very moment your eye absorbs a photon of light reflected from the page of this book, it emits a kind of antiphoton moving in the reverse direction that reaches the page just as the photon is emitted towards your eye. There is a 'handshake' between the page and your eye with connections going both ways in space and time.

Another way of looking at two-way flows in time in quantum mechanics was proposed by the quantum physicist Yakir Aharonov and his colleagues. Aharonov is best known as the co-predictor of the Aharonov-Bohm effect, a fundamental aspect of quantum theory related to superconductivity and several other quantum phenomena. Instead of the usual way of describing a quantum process as propagating only forwards in time, Aharonov and his colleagues also included quantum states propagating backwards: 'Time evolution is viewed as correlations between forward and backward states at adjacent moments.' Although most of their technical discussion concerned very short time scales, they pointed out a very radical implication if the same principles were applied to the universe as a whole. The final state of the universe – if there will be one – would work backwards, affecting events in the present:

> Quantum mechanics lets one impose a true future boundary condition – a putative final state of the universe. Philosophically or ideologically, one may or may not like the idea of a cosmic final state. This point is, however, that quantum mechanics offers a place to specify both an initial state *and* an independent final state. What the final state would be, if there is one, we don't know.[9]

Aharonov and his colleagues argued that time-reversed processes in quantum mechanics may be the tip of the iceberg of influences working 'backwards' in time.

But whether or not time-reserved processes occur within physical systems from actual futures, the influence of *virtual* futures or potentialities is of central importance in all developing patterns of organisation, including molecules.

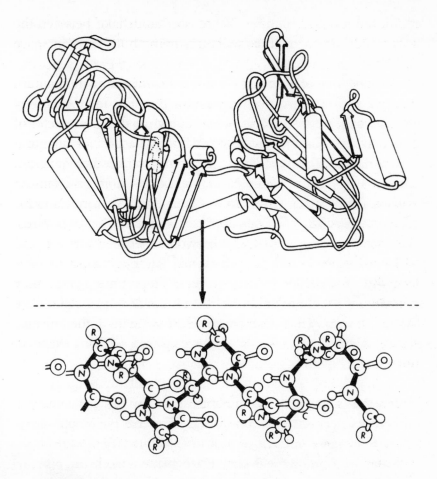

FIGURE 5.6 Above: The structure of the protein phosphoglycerate kinase, an enzyme isolated from horse muscle. The alpha-helices are represented by cylinders and the beta-strands by arrows. Below: the structure of part of an alpha helix in more detail, showing the relative positions of the atoms. (After Banks *et al.*, 1979)

Protein folding

The pull of processes towards attractors is not confined to living organisms. The formation of chemical molecules is also a type of morphogenesis; molecules are forms or structures. Their forms can be represented by attractors that lie at the bottom of potential wells (Figure 5.1): the molecules are stable because they are minimum-energy structures. If they are perturbed, and pushed away from the bottom of the well, they soon return to it.

With simple molecules like carbon dioxide, there is a simple, straightforward minimum energy structure. But with large, complicated molecules, like proteins, the range of possible structures becomes enormous. Protein molecules are made of polypeptide chains, strings of amino acids that twist, turn and fold into complex three-dimensional forms (Figure 5.6). A given type of protein molecule folds up into a unique structure. In the laboratory, many proteins can be made to unfold by changing their chemical environment; they then fold again correctly when they are replaced in appropriate conditions.[10] They return to a stable end-point.

This stable end-point is a minimum-energy structure at the bottom of a potential well. But this does not prove that it is the only structure with a minimum energy; there may be hundreds or thousands of other possible structures with the same minimum energy. Indeed, calculations to predict the three-dimensional structure of proteins, starting from the linear sequence of amino acids coded for by DNA, give far too many solutions.[11] In the literature on protein folding, this is known as the 'multiple-minimum problem'.[12]

There are persuasive reasons for thinking that the protein does not 'test' all these minima until it finds the right one. Christian Anfinsen, who won the Nobel Prize for his work on protein folding, put it thus:

If the chain explored all possible configurations at random by rotations about the various single bonds of the structure, it would take too long to reach the native configuration. For example, if the individual residues of an unfolded polypeptide chain can exist in only two states, which is a gross underestimate, then the number of possible randomly generated conformations is 10^{45} for a chain of 150 amino acid residues (although, of course, most of these would probably be sterically impossible ones). If each conformation could be explored with a frequency of a molecular rotation (10^{12} sec.$^{-1}$), which is an overestimate, it would take approximately 10^{26} years to examine all possible conformations.

Since the synthesis and folding of a protein chain such as that of ribonuclease or lysozyme can be accomplished in about 2 minutes, it is clear that all conformations are not traversed in the folding process. Instead, it appears to us that, in response to local interactions, the peptide chain is directed along a variety of possible low-energy pathways (relatively small in number), possibly passing through unique intermediate states, towards the conformation of lowest free energy.[13]

But not only may the folding process be 'directed' along certain pathways, it is also attracted towards one particular conformation of minimum energy, rather than any other possible conformations with the same minimum energy. The folding pathway can be thought of as a chreode in the morphogenetic field of the protein, and the final three-dimensional structure an attractor. Like biological morphogenesis, chemical morphogenesis is end-directed. Energy alone cannot select between these alternative possibilities and determine the specific structure taken up by the system.[14]

The failure of reductionism

Materialists used to believe that atoms were the ultimate eternal reality, and aspired to explain everything in terms of the physics and chemistry of these tiny particles and the interactions between them. They were the solid foundation on which all material explanations rested. But twentieth-century physics showed that atoms are not inert particles of solid stuff. They are structures of vibratory activity made up of subatomic particles, which are themselves vibratory patterns of activity. Reductionists now need to account for everything in terms of particle physics and the fundamental physical forces. Minds should be reducible to brains, brains to the chemistry and physics of nerve cells, cells to molecules, molecules to atoms and atoms to subatomic particles. In this atomistic spirit, many scientists are convinced that once physicists have explained the fundamental fields and particles everything else is nearly a matter of detail. Stephen Hawking expressed the standard view:

Since the structure of molecules and their reactions with each other underlie all chemistry and biology, quantum mechanics enables us in principle to predict nearly everything we see around us, within the limits set by the uncertainty principle. (In practice, however, the calculations required for systems containing more than a few electrons are so complicated that we cannot do them.)[15]

Even Lee Smolin, dissident though he may be about multiverse cosmology, is a conventional reductionist: 'Twelve particles and four forces are all we need to explain everything in the known world.'[16] Hawking and Smolin, like many other physicists, simply take it for granted that, given a comprehensive theory of the fundamental particles, all the phenomena of chemistry, life and mind will be explicable in terms of these microscopic entities. This is the old materialist agenda in a new guise. It is relatively easy to break things up and analyse the parts. The problem is to understand the whole; not just the parts but also their interactions need to be understood. And these interactions are not contained in the parts themselves. To study the molecules in a racing pigeon, it is necessary to kill the pigeon first, to grind up its tissues and cells, and separate the molecular components. But all the structure and activity of the pigeon has been destroyed in the process, just as the layout of a building is destroyed when it is demolished. The architecture of the building cannot be worked out from a chemical analysis of the rubble, nor can the form of the pigeon and its homing behaviour be reconstructed from an analysis of its molecules. Even if its genes are fully analysed and sequenced, it is not possible to predict the structure of the pigeon and the organisation of its behaviour, as discussed in the following chapter.

The reductionist approach ignores morphogenetic fields, chreodes and attractors. It assumes that everything can be worked out from the bottom up in terms of physical interactions and random collisions of particles, and from the past to the future. But this attempt is doomed to failure because of combinatorial explosions. One example is the failure of attempts to predict the three-dimensional structure of proteins by assuming that they explore all possible folding patterns at random until they find a stable minimum energy structure. As we have just seen, it would take a small protein about 10^{26} years to do this, far longer than the age of the universe, which is about 10^9 years. Moreover it would find no unique minimum energy structure because there are multiple minima.

As René Thom pointed out, the explanatory power of mathematics rapidly declines as systems become more complex:

> The excellent beginning made by quantum mechanics with the hydrogen atom peters out slowly in the sands of approximations in as much as we move towards more complex situations . . . This decline in the efficiency of mathematical algorithms accelerates when we go into chemistry. The interactions between two molecules of any degree of complexity evades precise mathematical description . . . In biology, if we make exceptions of the theory of population and of formal genetics, the use of mathematics is confined to modelling a few local situations (transmission of nerve impulses, blood flow in the arteries, etc.) of slight theoretical interest and limited practical value . . . The relatively rapid degeneration in the possible uses of mathematics when one moves from physics to biology is certainly known among specialists, but there is a reluctance to reveal it to the public at large . . . [T]he feeling of security given by the reductionist approach is in fact illusory.[17]

Thom argues that in modelling morphogenesis and behaviour, qualitative rather than quantitative mathematical models are needed, as in his models of morphogenetic fields, chreodes and attractors. Thom's models are topological, that is to say they are about forms, rather than about quantities. For example, in the capture chreode, an animal captures its prey, which is originally separate and external, and then ingests it. The prey is now inside the animal, and becomes part of it.[18]

Other modelling approaches include systems theory, which treats cells, organisms, societies or ecosystems as wholes with their own 'emergent properties', rather than trying to explain them from the bottom up. The parts of systems are related to each other through webs of relationships, including feedback loops.[19]

Thus there are three main holistic approaches. First, systems theorists aspire to build new kinds of mathematical models of systems' 'emergent properties', but implicitly assume that only known kinds of physical fields and forces are involved. Second, other holistic thinkers, like René Thom, are Platonists who look for ultimate explanations in mathematical forms or structures.[20] Third, there is the approach I follow myself: morphogenetic fields, chreodes and attractors are causal factors with properties that go beyond the familiar forces and fields of physics. They have time within them; they contain a memory of previous similar systems given by morphic resonance, and they attract organisms towards ends or goals through a kind of causation working 'backwards' in time. I discuss these ideas in more detail in the next chapter.

Are there purposes in evolution?

Does the evolutionary process as a whole have goals or attractors? Materialists say 'no' as a matter of principle. This denial is an inevitable historical consequence of the materialist philosophy.

The materialist denial of purposes in evolution is not based on evidence but is an assumption. Materialists are forced to attribute evolutionary creativity to chance on ideological grounds.

In the seventeenth century, the mechanistic revolution abolished souls and purposes from nature, with the single exception of human minds. Everything else, including human bodies, was explained mechanically in terms of pushing from the past, with no pull from the future. Nature was thought to go on like a machine indefinitely, made up of eternal matter in motion, following eternal laws. The only purposes were human and divine.

With the rise of materialism and atheism in the early nineteenth century, divine purposes were abolished, leaving only human purposes. And human purposes took on a new world-changing intensity as they were collectively funnelled into progress through science, technology and economic development. Most people still believed nature was fixed, although early evolutionary theories, like those of Erasmus Darwin and Lamarck, were pointing towards a different view.

With Charles Darwin's *Origin of Species* in 1859, biological evolution became mainstream. All life seemed to be engaged in a progressive development. Some scientists and philosophers thought that evolution showed the creativity of Nature herself; others, the imprint of divine creative activity; but atheists denied that there was any divine activity or purpose in evolution.

In the second half of the twentieth century, neo-Darwinians insisted that all creativity was in the final analysis a matter of random mutations and the blind forces of natural selection: an interplay of chance and necessity. And when the Big Bang theory became mainstream in the 1960s, the presuppositions of material-ism meant that the entire process of cosmic evolution must be purposeless, like biological evolution on earth.

Thus the standard scientific view is that both cosmic and biological evolution are purposeless. The fact that the universe is just right for life, at least on earth, as in the Anthropic Cosmological Principle, does not mean that the universe as a whole has any purpose. Among countless universes, it just happens to be the one that has the right conditions for life.

Gravitational attraction to the future

In models of attractors, as we have seen, gravity is the metaphor for attraction towards ends or goals – as in potential wells, dynam-ical attractors, attractors in morphogenetic fields, chreodes and the attractors of animal behaviour. All these models of purposive activity borrow their plausibility from our experience of gravity.

Gravitational attraction is so central to our experience that we take it for granted. We live and move and have our being in the field of gravity, like fish in water. If we drop things they fall. We walk upright and maintain our balance against the power of gravity. We succumb to it as we lie down to sleep. If we jump out of planes with parachutes at 30,000 feet, gravity carries us to earth.

Gravity is a force of attraction and pulls whatever is under its influence. An object in the gravitational field is pulled towards the future. Gravity attracts towards future ends. In this sense, it works backwards in time.

For a rock falling down a mountain, a gravitational pull from the future is not a metaphor but a description. But what about the evolution of the universe? Is everything being drawn towards a gravitational goal or attractor? The entire universe is within the universal gravitational field, which is not *in* space and time, but *is* space-time, according to Einstein's general theory of relativity. Gravity pulls everything together, and if opposing forces are not strong enough, it causes matter to collapse into black holes, as when heavy stars burn out. Likewise, if the energy that causes the universe to expand is less than a critical value, then the universe will begin to contract, and accelerate towards its end in the ultimate black hole, the Big Crunch. This is the final cosmic attractor, the end to which gravitation ultimately tends. And then perhaps it will give birth to a new universe.

Opposing the contractive pull of gravitation is the dark energy that makes space expand. If there is enough of it, according to Roger Penrose's theory (see Chapter 2), space will go on expanding exponentially until all structures will break down; matter will be diluted until all distinctions are lost in a featureless sea of photons and other massless particles.[21] For Penrose, this final state then somehow transforms itself into the Big Bang of the next universe.

In one scenario, everything is sucked into an ultimate black hole. Darkness triumphs. In the other, it is sublimated into infinite light. Light triumphs. Meanwhile, the contractive and expansive forces together sustain the universe. Expansive energy, pushing from the past, gives the universe an arrow of time, while through gravitation everything is pulled towards a future unity, at least a virtual unity, and maybe an actual unity as well.

All organisms within the universe are like scaled-down versions of this cosmic process: unifying fields pull them towards attractors in the future, and energy flowing from the past propels them forward. All are embedded within larger wholes – atoms in molecules, organelles in cells, animals in ecosystems, the earth within the solar system, the solar system within the galaxy – and all have their own ends and attractors.

Multiplicity and diversity

The unimaginably vast universe contains billions of galaxies, each with billions of stars. It extends beyond the limits of our ability to observe it, beyond the event horizon from which we can receive light or any other form of electromagnetic radiation. It contains countless atoms, molecules, crystals, stars and galaxies. On earth there is an immense diversity of living forms. In the human realm, there is a great variety of languages, cultural forms, social patterns, technical innovations, novels and films, sports, video games and so on. One essential feature of the universe seems to be fertility, multiplicity and creativity. Yet at the moment of the Big Bang there was none of this diversity. Multiplicity and diversity have increased through time, and so have complexities of organisation.

Materialists believe that this process can ultimately be explained in terms of energy, the laws of nature and chance, with no pulls from future ends of attractors. But this is an act of faith. They cannot prove that all evolution is purposeless; they assume it.

If there *are* purposes in evolution, one of them must be the proliferation of variety and complexity. Could creativity be an end in itself?

Some evolutionary philosophers, like Henri Bergson, saw the goal of the evolutionary process as ongoing creativity. Creativity is real; it is not the unfolding of a fixed plan. Bergson's God was a God who created himself through the evolutionary process: 'God has nothing of the already made; He is unceasing life, action, freedom. Creation, so conceived, is not a mystery; we experience it ourselves when we act freely.'[22] Underlying this creativity was what Bergson called 'the impetus of life' or the 'current of life'.

But the idea of ever-increasing complexity for ever, like the idea of an ever-expanding universe or an ever-expanding economy, is unsatisfying. We are used to stories with a beginning, middle and end.

Divine and human purposes

In the Judaeo-Christian tradition, human history is a journey with an end, and so is cosmic history. The beginning was the creation, when all was in harmony. Then came the Fall when Adam and Eve ate the fruit of the tree of knowledge of good and evil; the result was toil, suffering, competition, fighting and murder, as well as acts of goodness and prophecy; in other words, human history as we know it. Ultimately there is a culmination, a final redemption, a transformation. At the end of ordinary history Paradise will be restored and harmony re-established.

The protohistorical version of this story was the journey of the Jewish people out of bondage in Egypt, through the wilderness and to the Promised Land, where Paradise would be re-established on earth.

The reality was very different. When the Jewish people arrived in the Promised Land, it was not empty but inhabited by Palestinians. Then as now, endless conflict ensued. So the end of ordinary history was projected into the future, with the coming of the Messiah. For Christians, Jesus was the Messiah. But history still went on. Christian visionaries looked forward to a new end of history when Christ would come again and establish Paradise on earth for a thousand years.

Throughout the Middle Ages there was a succession of millenarian movements in Christian countries, well described by the historian Norman Cohn in his classic study *The Pursuit of the Millennium: Revolutionary Millenarians and Mystical Anarchists of the Middle Ages* (1957).[23] Francis Bacon, the first and greatest prophet of modern science, secularised this millenarian spirit. A new kind of journey to the Promised Land would be brought about by man himself, conquering nature. In the vanguard would be a scientific priesthood, whose purpose was 'the knowledge of causes, and secret motions of things; and the enlarging of human empire, to the effecting of all things possible'.[24] This vision of progress through science and technology became the basis of the secular philosophy of the Enlightenment. In its capitalist,

Communist and socialist forms it dominates almost all the modern world.

The discovery of the evolution of life in the nineteenth century and the evolution of the universe in the twentieth century placed human progress in a much wider context. But these discoveries also opened up a growing gulf between humanity and nature. Materialist science was pervaded by human purposes, not least by the desire for economic and technological progress, yet at the same time it denied the life and the purposes of nature. Many secular humanists believed that evolution somehow predicted – or even demanded – the continued upward development of humanity.[25] Meanwhile materialism in its economic and social manifestations has triumphed on a global scale. The effects on other species and on the earth's climate may be catastrophic.

The evolution of consciousness

All religions assume that human consciousness plays an essential role in the world and in human destiny. Humans have the potential to participate in ultimate Being, or God, or cosmic consciousness, or divine life, or nirvana. All religions started with a direct experience of this connection – through the ancient Indian seers or *rishis*, through the Buddha's enlightenment, through the Hebrew prophets, through Jesus Christ, and through Muhammad.

Experiences of unity with a greater being, or mystical experiences, are surprisingly common. The Religious Experience Research Unit at Oxford University, established in 1963 by the biologist Sir Alister Hardy, found that many thousands of people in Britain had felt they were 'in contact with a Being greater than themselves', and for most of these people, their mystical experiences had changed their lives. In addition, many thousands more had had near-death experiences, in most cases with life-changing effects.

Hinduism and Buddhism traditionally assume that lives and universes continue in endless cycles. They are repetitive rather than progressive. However, individual humans can escape by a kind of vertical take-off through establishing a connection with the universal mind or spirit.

Neither Hinduism nor the original forms of Buddhism are intrinsically evolutionary; indeed, in Hindu cosmology, in each cosmic cycle there are four ages, and we are currently in the last, the *kali yuga*, a time of strife and discord, when civilisation degenerates and people are as far as possible from God. By contrast, Tibetan Buddhists see a progressive process: enlightened beings come back in new incarnations to work for the liberation of all sentient beings. They will continue to do so until all have been liberated from the cycles of birth and death. And the Indian philosopher Sri Aurobindo (1872–1950) adopted a vision of spiritual as well as material evolution, and pointed towards a transformation of humanity, bringing about 'divine life on earth'.[26]

The Jesuit biologist Teilhard de Chardin (1881–1955) thought that the entire evolutionary process was moving towards an end-point of 'maximum organised complexity', which he called the Omega point. The Omega point was the attractor of the entire cosmic evolutionary process, and through it consciousness would be transformed.

Traditional religions grew up in a time when the known cosmos was small. Radio and space telescopes enable us to see far beyond our own galaxy to a vastly greater universe than anyone had ever imagined. If the transformation of human consciousness is the goal of evolution, then why do there need to be a billion stars beside our sun in our galaxy and billions of other galaxies beyond it? Is human consciousness unique? Or is consciousness developing throughout the universe? And will our consciousness ultimately make contact with those other minds? These are open questions. Neither conventional science nor traditional religions have any ready answers. Philosophers like Teilhard de Chardin and Sri Aurobindo point to new possibilities that go beyond the speculations of scientists by seeing consciousness as central to the evolutionary process. But even for the most materialistic of scientists, consciousness has a privileged position as the matrix of human knowledge, the basis of science itself.

What difference does it make?

At the personal level, a recognition of purposes in nature means that human purposes are not unique. Like animals and plants, our bodies have intrinsic powers to grow, heal and maintain themselves; we share goal-directed behaviour with other animals. Many of our goals, like capturing food, reproducing ourselves and behaving co-operatively with other members of our social groups, are similar to those of many other species. Our own lives, and those of our societies and cultures, are embedded in larger systems, like those of the earth, the solar system, the galaxy and ultimately the entire evolutionary universe. Without a wider sense of purpose our lives seem futile.

From a scientific point of view, the recognition of the purposes or goals of plants and animals opens up a deeper understanding than a mechanistic approach can offer.

A causal flow of influence from the virtual or even actual futures towards the present, from attractors towards the system they are attracting, has major implications for the understanding of nature in general, and of minds in particular. Influences from the future may even be detectable experimentally, as discussed in Chapter 9.

From a spiritual perspective, future connections with higher or more inclusive states of consciousness may serve as spiritual attractors, pulling individuals and communities towards experiences of higher unity.

Questions for materialists

How do you know that there are no purposes in nature? Is this merely an assumption?

If there are no purposes in nature, how can you have purposes yourself?

How do attractors attract?

Is there any evidence for the materialist belief that the entire evolutionary process is purposeless?

SUMMARY

Self-organising systems have their own ends or goals, attractors towards which they move. All living organisms show goal-directed development and behaviour. Developing plants and animals are attracted towards developmental ends, and if their development is disrupted they can often reach the same end by a different pathway. Animal behaviour is directed towards ends or 'consummatory acts'. In physics, goal-directed behaviour is modelled in terms of attractors, as if future ends had an influence working 'backwards' in time, and several quantum theorists have proposed that causal influences move from the future towards the past, as well as from the past towards the future. Chemical processes like protein folding also seem to be directed towards attractors or ends. End-directed behaviour is usually unconscious; even in humans, most purposes and goals are habitual. Conscious purposes are the exception rather than the rule. Both evolution and progress can be interpreted in terms of attractors, with influences working backwards in time from future goals.

6

Is All Biological Inheritance Material?

'Like father, like son' was a proverb in the Middle Ages; the Latin version, '*qualis pater talis filius*', played the same role in ancient Rome. The general principles of heredity had been known all over the world for millennia: children generally resemble their parents; they are usually more like members of their immediate family than unrelated people. It was also common knowledge that the same principles apply to animals and plants. Long before Darwin's theory of evolution and the pioneering genetic research of Gregor Mendel, people were breeding plants and animals selectively, creating an astonishing array of domesticated varieties, like dogs from Afghan hounds to Pekinese, and cabbages from broccoli to kale.

The discoveries of Mendel and Darwin were based on the practical successes of many generations of farmers and breeders. Darwin studied the subject for years. He subscribed to specialist publications like *Poultry Chronicle* and the *Gooseberry Growers' Register*, and grew fifty-four varieties of gooseberry in his garden at Down House, in Kent. He drew on the experience of cat and rabbit fanciers, horse and dog breeders, bee-keepers, horticulturalists and farmers. He joined two of the London pigeon clubs, visited fanciers to see their birds, and kept all the breeds he could obtain. He summarised this wealth of information in his book *The Variation of Animals and Plants Under Domestication* (1868), which is one of my favourite books on biology. The power of selective breeding suggested that a similar process worked spontaneously in the wild: natural selection.

Genetics is now at the very centre of biology. The standard view is that hereditary information is coded in the genes. The words 'hereditary' and 'genetic' are treated as synonyms. After the discovery of the structure of DNA in 1953, the nature of heredity appeared to be fully understood in molecular terms, at least in principle. The human genome project, completed in the year 2000, was a culminating technical triumph.

From a materialist point of view, non-material inheritance is impossible, except for cultural inheritance. Everyone agrees that cultural inheritance – say, through language – involves a transfer of information that is not genetic. But all other forms of inheritance *must* be material: there is no other possibility.

Several forms of material inheritance are known to be non-genetic. Cells inherit patterns of cell organisation and structures like mitochondria directly from their mother cells, not through genes in the cell nuclei. This non-nuclear inheritance is called cytoplasmic inheritance. Animals and plants are also influenced by characteristics acquired by their ancestors. An inheritance of acquired characteristics can take place *epigenetically*, as opposed to genetically, through chemical changes that do not affect the underlying genetic code, as discussed below.

I look first at the unfamiliar idea of the non-material transmission of form and organisation. This used to be the mainstream view; twentieth-century genetics developed in reaction against it. But even materialists end up with non-material explanations.

Immaterial forms

In the ancient world, almost no one believed that the form of an acanthus plant or a hawk was inherited through seeds or eggs alone. Platonists thought plants and animals were somehow shaped by the transcendent Idea or Form of their species. Modern Platonists, like René Thom, agree. They see the ideal Form of a species as a mathematical structure or model that is 'reified' in physical plants or animals. The mathematical model for an acanthus plant is not embedded in the genes: it exists in a mathematical realm that transcends space and time. Human mathematical models are mere approximations to these ultimate mathematical archetypes.

Aristotle, Plato's student, disagreed. The forms of the species were not outside space and time, but inside space and time. They were *immanent*, meaning 'dwelling in', not *transcendent*, meaning 'climbing beyond'. Instead of an archetype in a transcendent mind-like realm, the form of the body was in the soul, which attracted the developing animal or plant towards its final form (see pages 130–131). The soul served both as its formal cause, the cause of the body's form, and its final cause, the end or goal towards which the organism was attracted.

In the European Middle Ages, Aristotle's theory, as modified and interpreted by Thomas Aquinas, was the basis of the orthodox understanding of causation. A process of change, like the growth of a walnut tree from a nut, involved four kinds of cause. The material cause was the matter out of which the plant was made, the nut and the matter it took up from its surroundings as it grew, like water and minerals from the soil. The moving cause was the energy that powered it, from sunlight. The formal cause was the cause of the form or structure, the walnut-tree form in the plant's soul. The final cause was the goal or purpose of the plant's growth, namely the mature tree producing nuts to reproduce itself.

An architectural analogy provides another way of thinking about the four causes. In order to build a house, there must be building materials, like bricks and cement. These are the material causes. Putting them in the right places requires the energy of the builders and their machinery: these are the moving causes. The places in which they put the materials are specified by the architect's plan: this is the formal cause. All this activity is happening because the person paying for the house wants to live in it: this is the purpose or final cause. All four causes are necessary: the house would not exist without the materials of which it is made, or the energy of the build-ers, or a plan, or a motivation for building it. In living organisms, immaterial souls provide both plans and purposes.

An essential feature of the mechanistic revolution in the seven-teenth century was the abolition of souls, along with formal and final causes. Everything was to be explained mechanistically in terms of material and moving causes. This meant that the source of an organism's form must already be present inside the fertilised egg as a material structure.

Pre-formation and new formation

From the seventeenth century until the beginning of the twentieth, biologists were divided into two main camps: the mechanists and the vitalists. Both needed to explain heredity. The vitalists contin-

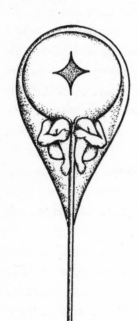

ued the Aristotelian tradition: organisms were shaped by souls or non-material vital forces. The problem was that they could not say how these non-material forces worked or how they interacted with bodies.

The mechanists preferred a material explanation, but they too soon ran into problems. To start with, they proposed that the animals and plants were already present in the fertilised egg in a miniature form. They were *pre-formed*. Development was a growth and unfolding – or inflation – of these pre-formed material structures. A few pre-formationists believed that the tiny unexpanded organisms came from eggs, but most thought they were in sperm, and some claimed to have proved it. One microscopist saw miniature horses in horse sperm, and miniature donkeys in donkey sperm, with big ears. Another saw tiny homunculi in human sperm (Figure 6.1).[1]

FIGURE 6.1 A human sperm containing a little man, or homunculus, as seen by a microscopist in the early eighteenth century. (After Cole, 1930)

Although pre-formationism was easy to understand and apparently supported by microscopic evidence, it ran into serious theoretical difficulties in relation to the succession of generations. As their vitalist opponents pointed out, if a rabbit grows from a miniature rabbit in a fertilised egg, the tiny rabbit in the egg must contain even tinier rabbits in its gonads, and so on *ad infinitum*.[2]

Pre-formationism was finally refuted in the late eighteenth century. As researchers looked at developing embryos in detail, they found that new structures appeared that were not there before. For example, the intestine, formed by the infolding of a sheet of tissue from the ventral surface, produced a gutter which in time transformed itself into a closed tube.[3] By the mid-nineteenth century the evidence was overwhelming: development involved the formation of new structures that were not already present. Development was *epigenetic*, from the Greek *epi*, over and above, and *genesis*, origination. New structures appeared that were not already present in the egg.

Epigenesis supported both Platonic and Aristotelian schools of thought. Neither supposed that all of an organism's form was contained in the matter of the fertilised egg. Its form was derived from a Platonic Idea or a soul.

By contrast, mechanists faced the daunting challenge of explaining how more material form could arise from less and develop in a highly ordered way. In the 1880s, August Weismann (1834–1914) thought he had found the answer. He made a theoretical division of organisms into two parts, the body, or somatoplasm, and the germ-plasm, a material structure present in the fertilised egg. He thought the germ-plasm was an active agency, containing 'determinants' that shaped the form of the somatoplasm. The germ-plasm affected the somatoplasm but not vice versa. The determinants 'directed' the formation of the adult organism, but the germ-plasm itself was passed on unchanged through eggs and sperm (Figure 6.1A).

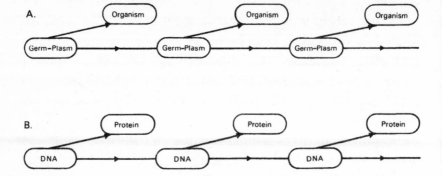

FIGURE 6.1A: Weismann's scheme for the continuity of the germ-plasm from generation to generation, with organisms as transient entities.

B: The 'central dogma' of molecular biology in which Weismann's scheme is interpreted in terms of DNA and proteins.

By the middle of the twentieth century, the discovery of genes located in chromosomes inside cell nuclei seemed to have confirmed Weismann's theory. The genes were the germ-plasm, replicated more or less unchanged in every cell division. The discovery of the structure of the genetic material, DNA, and the cracking of the genetic code in the 1950s showed how the Weismann doctrine could be reduced to the molecular level. DNA was the germ-plasm, proteins were the somatoplasm (Figure 6.1B). The DNA coded for the structure of proteins, but not vice versa; Francis Crick called this the 'central dogma' of molecular biology. Meanwhile, the neo-Darwinian theory explained evolution in terms of random mutations in genes and changes in gene frequencies in populations as a result of natural selection. The triumphs of molecular genetics combined with the neo-Darwinian theory of evolution seemed to provide overwhelming evidence for the material theory of inheritance. But this triumph was more a matter of rhetoric than reality.

Why genes are overrated

There is a vast gulf between rhetoric about the powers of genes and what they actually do. Investors in biotechnology are swept along by the metaphors, as are readers of popular science. The problem goes right back to Weismann, who made the determinants an active agency, controlling and directing the organism's development. In effect, he endowed a special kind of matter, the germ-plasm, with the properties of the soul. Genetic programs and selfish genes are similarly endowed with vital powers, including the ability to 'mould matter' and 'create form'.[4]

Thanks to the discoveries of molecular biology, we know what genes actually do. They code for the sequences of amino acids that are strung together in polypeptide chains, which then fold up into protein molecules. Also, some genes are involved in the control of protein synthesis.

DNA molecules are molecules. They are not 'determinants' of particular structures, even though biologists often speak of genes 'for' structures or activities, such as genes 'for' curly hair or 'for' nest-building behaviour in sparrows. Genes are not selfish and ruthless, as if they contained gangster homunculi. Nor are they plans or instructions for organisms. They merely code for the sequences of amino acids in protein molecules.

Richard Dawkins has probably done more than any other author to popularise genes. Unfortunately his vivid metaphors are highly misleading. For example, this is how he describes how all the cells of the human body contain copies of the complete set of human DNA:

> This DNA can be regarded as a set of instructions for how to make a body . . . It is as though, in every room of a gigantic building, there was a book-case containing the architect's plans for the entire building. The 'book-case' in a cell is called the nucleus. The architect's plans run to 46 volumes in humans – the number is different in other species. The 'volumes' are called chromosomes.[5]

What Dawkins does is to project on to the DNA molecules the purposive vital factors of vitalism, trying to squeeze the soul into chemical genes, which are thereby endowed with instructions, plans, purposes and intentions they cannot possibly have. He admits that these are metaphors, adding, 'Incidentally, there is of course no "architect".'[6] But despite occasional disclaimers, the entire force of his argument depends on anthropocentric metaphors and molecules that have come to life. He is a vitalist in molecular clothing.

The metaphor of the genetic program is another kind of cryptovitalism. The purposive vital factor is a computer program. This metaphor attempts to bridge the gulf between hereditary characteristics – say, the form of a sunflower – and the DNA and protein molecules within it. If the genes somehow *program* the development of the sunflower, then the gulf between this complex living structure and the DNA molecules within it seems less disturbing, even though almost nothing is known about the nature of the sunflower program and how it produces a sunflower.

The metaphor of the genetic program inevitably suggests that development is organised by a pre-existing purposive principle that is either mind-like or designed by a mind. Computer programs are intelligently designed by human minds for particular purposes, and act upon and through the electronic machinery of a computer. The computer is a machine, but the program is not.

Significantly, the analogy between programs and souls played a central role in the thinking of one of the founders of modern computational theory, Alan Turing. As a young man, he was intensely preoccupied with the question of survival, following the death of his beloved friend Christopher Morcomb in 1930. At first he adopted a traditional dualistic view, arguing in favour of a non-material spirit. He later found a more scientific-sounding model of the mind as a system of programs. Such programs could be 'embodied' in particular physical machines but were themselves independent of material incarnation.[7] The program could survive the destruction of any particular computer, and be embodied in another one, like a transmigrating soul.

If genetic programs were carried in the genes, then all the cells of the body would be programmed identically, because in general they contain exactly the same genes. The cells of your arms and legs, for example, are genetically identical. Your limbs contain exactly the same kinds of protein molecules, as well as chemically identical bone, cartilage and nerves. Yet arms and legs have different shapes. Clearly, the genes alone cannot explain these differences. They must depend on formative influences that act differently in different organs and tissues as they develop. These influences cannot be inside the genes: they extend over entire tissues and organs. At this stage, in most conventional explanations, the concept of the genetic program fades out, and is replaced by vague statements about 'complex spatio-temporal patterns of physico-chemical activity not yet fully understood' or 'mechanisms as yet obscure' or 'chains of parallel and successive operations that build complexity'.[8]

In spite of the fact that many biologists now recognise that it is misleading, the genetic program continues to play a large conceptual role in modern biology. There seems to be a need for such an idea. Mechanistic biology grew up in opposition to vitalism. It defined itself by denying that living organisms are organised by purposive, mind-like principles,[9] but then reinvented them in the guise of genetic programs and selfish genes. The dominant paradigm of modern biology, although nominally mechanistic, is remarkably similar to vitalism, with 'programs' or 'information' or 'instructions' or 'messages' playing the role formerly attributed to souls.

Mechanists have always accused vitalists of trying to explain the mysteries of life in terms of empty words, such as vital factors and souls, that 'explain everything and therefore nothing'. But the vital factors in their mechanistic guises have exactly this characteristic. How does a marigold grow from a seed? Because it is genetically programmed to do so. How does a spider instinctively spin its web? Because of the information coded in its genes. And so on.

The unfulfilled promises of molecular biology

It is hard to recall the atmosphere of exhilaration in the 1980s as new techniques enabled genes to be cloned and the sequence of 'letters' in their genetic code to be discovered. This seemed like biology's crowning moment: the genetic instructions of life itself were finally laid bare, opening up the possibility for biologists to modify plants and animals genetically, and grow richer than they could ever have imagined. Almost every week newspaper head-lines reported a new breakthrough: 'Scientists find genes to combat cancer', 'Gene therapy offers hope to victims of arthritis', 'Scientists find secret of ageing', and so on.

The new genetics seemed so promising that soon the entire spectrum of biological researchers was busy applying its tech-niques to their specialities. Their remarkable progress led to a vast, ambitious vision: to spell out the full complement of genes in the human genome. As Walter Gilbert of Harvard University put it, 'The search for this "Holy Grail" of who we are has now reached its culminating phase. The ultimate goal is the acquisition of all the details of our genome.' The Human Genome Project was formally launched in 1990 with a projected budget of $3 billion.

The Human Genome Project was a deliberate attempt to bring 'Big Science' to biology, which had previously been more like a cottage industry. Physicists were used to huge budgets, partly as a result of the Cold War: there was enormous expenditure on missiles and hydrogen bombs, Star Wars, multi-billion-dollar particle accelerators, the space programme and the Hubble Space Telescope. Ambitious biologists suffered from physics envy. They dreamed of the days when biology would have high profile, high prestige and multi-billion-dollar projects. The Human Genome Project was the answer.

At the same time, a tide of market speculation in the 1990s led to a boom in biotechnology, reaching a peak in 2000. In addition to the official Human Genome Project, Celera Genomics carried out a private genome project, headed by Craig Venter. The company planned to patent hundreds of human genes and own the commercial rights to them. Celera Genomics' market value, like that of many other biotechnology companies, rocketed to dizzy heights in the early months of 2000.

Ironically, the rivalry between the public and private genome projects led to a bursting of the bubble before the sequencing of the genome had even been completed. In March 2000, the leaders of the public genome project publicised the fact that all their information would be freely available to everyone. This led to a statement by US President Clinton on 14 March 2000: 'Our genome, the book in which all human life is written, belongs to every member of the human race ... We must ensure that the profits of the human genome research are measured not in dollars, but in the betterment of human life.'[10] The press reported that the president planned to restrict genomic patents. The stock markets reacted dramatically. In Venter's words, there was a 'sickening slump'. Within two days, Celera's valuation lost $6 billion, and the wider market in biotechnology shares collapsed by $500 billion.[11]

In response to this crisis, a day after his speech President Clinton issued a correction, saying that his statement had not been intended to have any effect on the patentability of genes or the biotechnology industry. But the damage had been done. The stock-market valuations never recovered. And although many human genes were subsequently patented, very few proved profitable to the companies that owned them.[12]

On 26 June 2000, President Clinton and the British prime minister, Tony Blair, with Craig Venter and Francis Collins, the head of the official project, announced the publication of the first draft of the human genome. At the press conference in the White House, President Clinton said, 'We are here today to celebrate the completion of the first survey of the entire human genome. Without a doubt this is the most important, most wondrous map ever produced by mankind. It will revolutionise the diagnosis, prevention and treatment of most, if not all, human diseases ... Humankind is on the verge of gaining immense, new power to heal.' The British science minister, Lord Sainsbury, said, 'We now have the possibility of achieving all we ever hoped for from medicine.'[13] One of the editors of *Nature* proclaimed that by the end of the twenty-first century, 'genomics will allow us to alter entire organisms out of recognition, to suit our needs and tastes ... [and] will allow us to fashion the human form into any conceivable shape. We will have extra limbs, if we want them, and maybe even wings to fly.'[14]

This astonishing achievement of sequencing the human genome has indeed transformed our view of ourselves, but not as anticipated. The first surprise was that there were so few genes. Rather than the predicted 100,000 or more, the final tally of about 23,000 was very puzzling, and all the more so when compared with the genomes of other animals much simpler than ourselves. There are about 17,000 genes in a fruit fly, and about 26,000 in a sea urchin. Many species of plants have far more genes than us – rice has about 38,000, for example.

In 2001, the director of the chimpanzee genome project, Svante Paabo, anticipated that when the sequencing of the ape's genome was completed, it would be possible to identify 'the profoundly interesting genetic prerequisites that make us different from other animals'. When the complete chimpanzee sequence was published four years later, his interpretation was more muted: 'We cannot see in this why we are so different from chimpanzees.'[15]

The 'missing heritability problem'

In the wake of the Human Genome Project, the mood changed dramatically. The optimism that life would be understood if molecular biologists knew the 'programs' of an organism gave way to the realisation that there is a huge gap between gene sequences and actual human beings. In practice, the predictive value of human genomes turned out to be small, in some cases less than that achieved with a measuring tape. Tall parents tend to have tall children, and short parents short children. By measuring the height of parents, their children's heights can be predicted with 80 to 90 per cent accuracy. In other words, height is 80 to 90 per cent heritable. Recent 'genome-wide association studies' compared the genomes of 30,000 people and identified about fifty genes associated with tallness or shortness. To everyone's surprise, taken together, these genes accounted for only about five per cent of the inheritance of height. In other words, the 'height' genes did *not* account for 75 to 85 per cent of the heritability of height. Most of the heritability was missing. Many other examples of missing heritability are now known, including the heritability of many diseases, making 'personal genomics' of very questionable value. Since 2008, in scientific literature this phenomenon has been called the 'missing heritability problem'.

In 2009, twenty-seven prominent geneticists, including Francis Collins, the former head of the Human Genome Project, published a paper in *Nature* on the missing heritability of complex diseases in which they acknowledged that, despite more than seven hundred genome-scanning publications and an expense of more than $100 billion, geneticists had found only a very limited genetic basis for human diseases.[16] In 2010, in a special series of articles in *Nature* to celebrate the tenth anniversary of the completion of the first draft of the human genome, a common theme was the 'mismatch' between the sophistication of the data collection and understanding it. In an article called 'A reality check for personalised medicine', the authors observed, 'Never before has the gap between the quantity of information and our ability to interpret it been so great.'[17]

In 2011, to celebrate the tenth anniversary of the actual publication of the human genome, the tone was even more modest: 'Although genomics has already begun to improve diagnostics and treatments in a few circumstances, profound improvements in the effectiveness of healthcare cannot realistically be expected for many years.'[18] Some critics go further. Jonathan Latham, director of the Bioscience Resource Project, commented,

The most likely explanation for why genes for common diseases have not been found is that, with few exceptions, they do not exist . . . The likelihood that further searching might rescue the day appears slim. A much better use of the money would be to ask: if inherited genes are not to blame for our commonest illnesses, can we find out what is?[19]

Meanwhile, the optimism of stock-market investors has suffered recurring blows. After the biotech bubble burst in 2000, many biotech companies either went out of business or were taken over by pharmaceutical or chemical corporations. An article in the *Wall Street Journal* in 2004 entitled 'Biotech's Dismal Bottom Line: More than $40 billion in Losses'[20] went on to say, 'Biotechnology . . . may yet turn into an engine for economic growth and cure deadly diseases. But it's hard to argue that it's a good investment. Not only has the biotech industry yielded negative financial returns for decades, it generally digs its hole deeper every year.'[21]

In 2006 Harvard Business School published a detailed analysis of the industry. They found that 'only a very tiny fraction' of biotechnology companies had ever made a profit, and that promises of breakthroughs had failed over and over again. Defenders of the industry argued that more time was needed, but the Harvard Business School analysis pointed to the opposite conclusion: '[G]iven the extremely poor long-term performance of the biotechnology industry in general, and specific firms in particular, capital has been, if anything, too patient.'[22]

Despite its disastrous business record, this vast investment in molecular biology and biotechnology has had wide-ranging effects on the practice of biology, if only by creating so many jobs. The demand for graduates in molecular biology has transformed the teaching of biology. The molecular approach now predominates in most universities and it has strongly influenced science teaching in secondary schools.

Precisely because there has been such a strong emphasis on molecular biology, its limitations are becoming increasingly apparent. The sequencing of the genomes of ever more species of animals and plants, together with the determination of the structures of thousands of proteins, is causing molecular biologists to drown in their own data. There is practically no limit to how many more genomes they can sequence or proteins they can analyse. Molecular biologists now rely on computer specialists in the rapidly growing field of bioinformatics to store and try to make sense of this unprecedented quantity of information, sometimes called the 'data avalanche'.[23] What does it all mean?

The advances of molecular biology have led to other big surprises. In the 1980s, there was great excitement when a family of genes called homeobox genes was discovered in fruit flies. Homeobox genes determine where limbs and other body segments will form in a developing embryo or larva; they seem to control the pattern in which different parts of the body develop. Mutations in these genes can lead to the growth of extra, non-functional body parts.[24] These are called homeotic mutations, as discussed below. At first sight, the homeobox appeared to provide the basis for a molecular explanation of morphogenesis: here were the key switches. At the molecular level, homeobox genes act as templates for proteins that 'switch on' cascades of other genes.

This study of genes involved in the regulation of development is part of a growing field called evolutionary developmental biology, or evo-devo for short. But here too, molecular biology is a victim of its own success: it has shown that morphogenesis itself continues to elude a molecular explanation. The molecular control systems turned out to be very similar in widely different animals. Homeobox genes are almost identical in flies, reptiles, mice and humans. Although they play a role in the determination of the body plan, they cannot explain the organisms' shapes. Since the genes are so similar in fruit flies and in us, they cannot explain the differences between flies and humans. It was shocking to find that the diversity of body plans across many different animal groups was not reflected in diversity at the level of the genes. As some leading molecular biologists commented, 'Where we most expect to find variation, we find conservation, a lack of change.'[25]

The genome wager

By 2009, it was clear that many of the promises of the genome project had not been fulfilled. But many biologists still believed that the genome in principle explained the organism. For example, Lewis Wolpert, an eminent British biologist, gave a ringing statement of faith in the role of the genes and their explanatory power when he proclaimed that, with more information and enormous computing power, 'We would, given a fertilised human egg, be able to have a picture of all the details of the newborn baby, including any abnormalities. We would also be able to programme the egg to develop into any shape we desire. The time will come when this is possible.'[26]

A few months later, Wolpert and I met to debate 'The Nature of Life' as the final event of the 2009 Cambridge University Science Festival.[27] Wolpert reaffirmed his faith in the predictive power of the genome, and I challenged him to a bet. I said I was prepared to bet that his prediction would not come true in ten years, or even twenty. After a few moments' consideration, he said it might take a hundred years. This was clearly an unverifiable prediction for anyone alive today. After our public debate, we continued our discussion, and I asked him what he thought could be achieved in twenty years. At first he thought that all the details of a mouse might be predictable on the basis of its genome. Then, after further consideration, he scaled down his prediction from mice to chickens, then frogs, then nematode worms. We finally agreed on a formal wager, published in *New Scientist* in July 2009.[28] At stake is a case of fine port, Quinta do Vesuvio 2005, for which we paid half each, and which is being stored in the cellars of the Wine Society, near London. Experts say it should be perfectly mature in 2029. The wager is:

By 1 May 2029, given the genome of a fertilised egg of an animal or plant, we will be able to predict in at least one case all the details of the organism that develops from it, including any abnormalities.

Wolpert bets that this will happen. I bet that it will not. If the outcome is not obvious, the Royal Society will be asked to adjudicate.

I think Wolpert's faith in the predictive power of the genome is misplaced because genes enable organisms to make proteins, but do not explain the development of embryos. The problems begin with the proteins themselves. Genes code for the linear sequences of amino acids in proteins, which then fold up into complex three-dimensional forms. Wolpert presupposes that, given the sequence of amino acids specified by the genes, the folding of proteins can be computed from first principles. This has proved impossible, despite more than forty years of intensive, well-funded research (see Chapter 5). Even if the protein-folding problem could be solved, the next stage would be to attempt to predict the structures of cells on the basis of the interactions of hundreds of millions of proteins and other molecules, unleashing a vast combinatorial explosion, with more possible arrangements than all the atoms in the universe.

Random molecular permutations simply cannot explain how organisms work. Instead, cells, tissues and organs develop in a modular manner, shaped by morphogenetic fields, first recognised by developmental biologists in the 1920s (see Chapter 5). Wolpert himself acknowledges the importance of such fields. Among biologists, he is best known for his concept of 'positional information', by which cells 'know' where they are within the morphogenetic field of a developing organ, such as a limb. But he believes that morphogenetic fields can be reduced to standard chemistry and physics. I disagree. I propose that these fields have organising abilities or systems properties that involve new scientific principles.

The predictive powers of the genome have been reduced yet further by the recognition of epigenetic inheritance.

Epigenetics and the inheritance of acquired characteristics

One of the biggest controversies in twentieth-century biology concerned the inheritance of acquired characteristics, the ability of animals and plants to inherit adaptations acquired by their ancestors. For example, if a body-builder acquired enormous muscles, his children would tend to have larger muscles as a result. The opposing view, promoted by August Weismann (Figure 6.1) and by the science of genetics, denied that organisms could inherit features their ancestors had acquired; they only passed on 'determinants' or genes that they had themselves inherited.

In Darwin's day, most people assumed that acquired characteristics could indeed be inherited. Jean-Baptiste Lamarck had taken this for granted in his theory of evolution published more than fifty years before Darwin's, and the inheritance of acquired characters was often referred to as 'Lamarckian inheritance'. Darwin shared this belief and cited many examples to support it.[29] In this respect Darwin was a Lamarckian, not so much because of Lamarck's influence but because he and Lamarck both accepted the inheritance of acquired characteristics as a matter of common sense.[30]

Lamarck placed a strong emphasis on the role of behaviour in evolution: animals' development of new habits in response to needs led to the use or disuse of organs, which were accordingly either strengthened or weakened. Over a period of generations, this process led to structural changes that became increasingly hereditary. Lamarck's most famous example was the giraffe, whose long neck was acquired through the habit of stretching up to eat the leaves of trees over many generations (see Chapter 1). In this respect too, Darwin agreed with Lamarck, and he provided various illustrations of the hereditary effects of the habits of life. For example, ostriches, he suggested, may have lost the power of flight through disuse and gained stronger legs through increased use over successive generations.[31] Darwin was very conscious of the power of habit, which was for him almost another name for nature. Francis Huxley summarised Darwin's attitude:

A structure to him meant a habit, and a habit implied not only an internal need but outer forces to which, for good or evil, the organism had to become habituated . . . In one sense, therefore, he might well have called his book *The Origin of Habits* rather than *The Origin of Species*.[32]

The problem was that no one knew how acquired characteristics could be inherited. Darwin tried to explain it with his hypothesis of 'pangenesis'. He proposed that all the units of the body threw off tiny 'gemmules' of 'formative matter', which were dispersed throughout the body and aggregated in the buds of plants and in the germ cells of animals, through which they were transmitted to the offspring.[33]

The neo-Darwinian theory of evolution, which became ortho-dox in the West in the twentieth century, differed from the Darwinian theory in denying the inheritance of acquired charac-teristics in favour of genes. Lamarckian inheritance was treated as a heresy. By contrast, in the Soviet Union the inheritance of acquired characteristics became the orthodox doctrine from the 1930s to the 1960s. Under the leadership of Trofim Lysenko, much research seemed to support the inheritance of acquired character-istics. Lysenko's power was backed by Stalin, and Mendelian geneticists were persecuted, and some were killed,[34] which further increased opposition to the inheritance of acquired characteristics in the West. The scientific question about the nature of inherit-ance became so intensely politicised that ideology, rather than scientific evidence, dominated the dispute.

The taboo against the inheritance of acquired characteristics began to dissolve around the turn of the millennium. There is a growing recognition that some acquired characteristics can indeed be inherited. This kind of inheritance is now called 'epigenetic inheritance'. In this context, the word 'epigenetic' signifies 'over and above genetics'. Some kinds of epigenetic inheritance depend on chemical attachments to genes, particularly of methyl groups. Genes can be 'switched off' by the methylation of the DNA itself or of the proteins that bind to it.

This is a fast-growing field of research, and there are many examples of epigenetic inheritance in plants and animals. For example, the effects of toxins can echo for generations. In one study, when pregnant rats were exposed to a commonly used agri-cultural fungicide, the development of their sons' testes was impaired, and they had a low sperm count later in life. *Their* sons also had lower sperm counts, and this effect was passed on from fathers to sons for four generations.[35] The inheritance of acquired characteristics occurs in invertebrates, like *Daphnia*, the water flea. When predators are around, the water fleas develop large defensive spines. When they reproduce, their offspring also have these spines even if they are not exposed to predators.[36]

Epigenetic inheritance also occurs in humans. A study in Sweden of men born between 1890 and 1920 showed that their nutrition in childhood affected the incidence of diabetes and heart disease in their grandchildren. Many common diseases that are inherited within families can also be passed on epigenetically.[37] The Human Epigenome Project, an international public-private consortium, was launched in 2003 to help to co-ordinate research in this rapidly growing field of enquiry.[38]

Although epigenetic inheritance breaks the taboo against the inheritance of acquired characteristics, it does not challenge the materialist assumption that heredity is material; it is another kind of material inheritance. It affects which genes are 'switched on' or 'switched off', and consequently affects what proteins a cell makes. But genes and proteins cannot in themselves explain morphogenesis or instinctive behaviour.

Morphic resonance and morphogenetic fields

The only way of making sense of inherited patterns of organisation is in terms of top-down causation by higher-level patterns, or 'systems properties', or fields.

One way of grasping how top-down causation works through fields is to think of the field of a magnet. Influences flow 'up' and 'down', to and from the overall field. The field of the magnet as a whole emerges from the lining up of the small magnetic domains within it. The field in turn works back on these constituent domains and keeps them lined up. If a magnet is heated above a critical temperature, it loses its magnetism; the order breaks down, and the microscopic magnetic domains are oriented randomly. The overall field is lost. This is like an organism dying.

Morphogenetic fields contain a nested hierarchy of morphogenetic units or holons (see Chapter 1, Figure 1.1). The morphogenetic field of a lemur co-ordinates the fields of its limbs, muscles and organs, the fields of organs those of tissues; those of the tissues co-ordinate the fields of the cells, and so on.

There are two main ways of thinking about morphogenetic fields. The first is to treat them as essentially mathematical structures, in which case we arrive again at a Platonic theory of form, as René Thom made explicit. The inheritance of form then becomes a matter of chemical genes and proteins interacting with timeless mathematics. The genes and proteins do not provide the form; the maths does.

Alternatively, morphogenetic fields may have history within them. They inherit their forms by morphic resonance from previous similar organisms. They can still be modelled mathematically, but these models do not *explain* the fields, they just model them. Inheritance depends on genes *and* on morphic resonance.

The difference between the Platonic theory and the morphic-resonance hypothesis can be illustrated by analogy with a television set. The pictures on the screen depend on the material components of the set and the energy that powers it, and also on the invisible transmissions it receives through the electromagnetic field. A sceptic who rejected the idea of invisible influences might try to explain everything about the pictures and sounds in terms of the components of the set – the wires, transistors, and so on – and the electrical interactions between them. Through careful research he would find that damaging or removing some of these components affected the pictures or sounds the set produced, and did so in a repeatable, predictable way. This discovery would reinforce his materialist belief. He would be unable to explain exactly how the set produced the pictures and sounds, but he would hope that a more detailed analysis of the components and more complex mathematical models of their interactions would eventually provide the answer.

Some mutations in the components – for example, by a defect in some of the transistors – affect the pictures by changing their colours or distorting their shapes; while mutations of components in the tuning circuit cause the set to jump from one channel to another, leading to a completely different set of sounds and pictures. But this does not prove that the evening news report is produced by interactions among the TV set's components. Likewise, genetic mutations may affect an animal's form and behaviour, but this does not prove that form and behaviour are programmed in the genes. They are inherited by morphic resonance, an invisible influence on the organism coming from outside it, just as TV sets are resonantly tuned to transmissions that originate elsewhere.

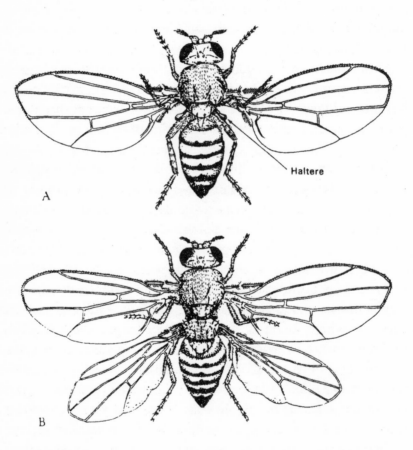

FIGURE 6.2A: A normal fruit fly. B: A mutant fly in which the third thoracic segment has been transformed into a duplicate of the second thoracic segment, bearing a pair of wings rather than halteres. Such flies are called bithorax mutants.

Some genetic mutations affect the tuning, with the result that a part of the embryo resonates with one morphogenetic field rather than another, resulting in a different structure, like a TV set tuned to a different channel. For example, fruit flies, like other flies, normally have two wings, and behind them a pair of balancing organs called halteres (Figure 6.2 A). Mutations in particular genes (in the *bithorax* gene complex) can cause an extra pair of wings to develop instead of halteres (Figure 6.2 B). Mutations of this kind are called homeotic mutations. Another kind of homeotic mutation in fruit flies causes legs to develop instead of antennae. In plants, homeotic mutations likewise cause some structures to be replaced by others – for example, in pea plants, one kind of homeotic mutant produces leaves with no tendrils: all the tendrils are replaced by leaflets. In another mutant, all the leaflets are replaced by tendrils. This does not mean that the altered genes 'program' leaflets or tendrils, or legs or antennae. Instead it means that the mutant genes affect the tuning system whereby the embryonic structures that normally tune into antenna fields tune into leg fields instead, or tendril fields instead of leaflet fields.

Other kinds of mutants affect the *details* of structures, just as some defects in the components of a television set affect the details of the sounds or pictures. For example, some mutant flies have white eyes instead of the normal red eyes. A mutation in a gene that codes for an enzyme that helps synthesise the red eye pigment results in flies that are unable to make the red pigment, so their eyes are white. There is a satisfyingly simple explanation: a randomly mutated gene gives rise to a defective enzyme, resulting in a change in eye colour. But this detail does nothing to explain the morphogenesis of the eye itself, organised by a nested hierarchy of chreodes in morphogenetic fields, pulled towards their morphogenetic attractors, namely mature, functioning eyes.

Platonists hope that one day these fields will be explained mathematically. The only real alternative is that morphogenetic fields are inherited by morphic resonance from previous similar organisms, together with their chreodes and attractors. This inheritance is not material, but it is nevertheless physical in the sense that it is natural, not supernatural. It involves a resonant transfer or form, or in-form-ation, from the past to the present. This memory-resonance from the past takes place through or across time and space. It does not fall off with distance, but works on the basis of similarity: the more similar, the more resonant.

The hypothesis of morphic resonance is testable experimentally. If fruit flies develop abnormally under abnormal conditions, then the more the abnormality occurs, the more likely it will be to happen again under the same conditions, through cumulative morphic resonance. And if animals, like squirrels, learn a new trick in one place, the more squirrels that learn it, the easier it should become for squirrels of the same species all over the world. There is already experimental evidence that these effects occur, discussed in detail in my books *A New Science of Life* and *The Presence of the Past*.

Twins

The relative importance of nature and nurture or genes and environment is not only a scientific but also a political question. From the nineteenth century onwards it stirred up intense passions. The liberal philosopher John Stuart Mill (1806–73) preached a gospel of social progress, whereby political and economic reforms would change human nature through changing the environment, ideas that had a strong influence on progressive political movements like liberalism, socialism and Communism.

On the other hand, Francis Galton, Charles Darwin's cousin, made a strong scientific case for the predominance of heredity, which is often taken to support a more conservative political philosophy. In his book *Hereditary Genius* (1869), he argued that the prominence of Britain's most distinguished families depended more on nature than on nurture. Galton was a pioneering advocate of eugenics, a word he coined. He also realised that the question of nature and nurture could be studied with the help of identical twins. He argued that identical twins had a similar heredity constitution, while non-identical twins were no more similar than ordinary brothers and sisters. And, sure enough, he found remarkable similarities between identical twins in a wide range of characteristics including the onset of disease and even time of death.[39]

Some political philosophers used Galton's ideas on heredity to justify the British class system, and Galton himself proposed that the state should regulate the fertility of the population in such a way as to favour the improvement of human nature through selective breeding. The eugenics movement had a large following in the United States and reached its apogee in Nazi Germany. Not surprisingly, Nazi scientists were very interested in twins. The notorious Josef Mengele's favourite project at the Auschwitz death camp was a study of identical twins, who were kept in special barracks. Mengele told one of his colleagues, 'It would be a sin, a crime . . . not to utilise the possibilities that Auschwitz had for twin research. There would never be another chance like it.'[40]

Meanwhile, psychologists of the behaviourist school took the opposite approach. They believed in promoting human progress through environmental conditioning. As John B. Watson, the founder of behaviourism, expressed it,

> Suppose we were to take individual twins into the laboratory and begin rigidly to condition them from birth to the twentieth year along utterly different lines. We might even condition one of the children to grow up without language. Those of us who have spent years in the conditioning of children and animals cannot help but realise that the two end products would be as different as day is from night.[41]

After the Second World War, the leading figure in identical-twin research was the educational psychologist Sir Cyril Burt, who claimed to have studied fifty-three pairs who had been reared apart. Burt focused on the inheritance of intelligence, as measured by IQ tests, and claimed that genetics had a far stronger influence than environment. Unfortunately for genetic determinists, Burt was shown to have faked some of his data.[42] But with the cracking of the genetic code and the growth of molecular biology in the

1960s, genetic determinism became increasingly influential, and was reinforced by new studies of twins, most notably in the Minnesota Twin Study, established in 1989.

The team at the University of Minnesota studied 1,400 pairs of identical twins and fraternal twins, including twins separated soon after birth. They found that separated identical twins showed remarkable similarities in a variety of characteristics, such as sense of wellbeing, social dominance, alienation, aggression and achievement. They also found a high correlation of IQ, almost identical to the figure that Burt was accused of inventing.[43] And there were some exceptionally striking similarities. For example, the 'Jim' twins (both called James by their adoptive families), who were separated soon after birth, showed remarkable similarities in their life histories. Both lived in the only house on the block, with a white bench around a back tree in the backyard; both were interested in stock-car racing; both had elaborate workshops where they made miniature picnic tables or miniature rocking chairs.[44] They also had similar health histories.[45]

Morphic resonance casts a new light on the studies of monozygotic twins. Because they are genetically identical and share the same womb throughout their embryonic development, they are much more similar to each other than any other pairs of humans. The greater the similarity, the stronger the morphic resonance. Hence the morphic resonance between identical twins will be stronger than between any other people. As a result, their patterns of activity, habits or health issues are likely to influence them by morphic resonance, even if they were separated soon after birth. Many of the remarkable similarities between identical twins may depend on morphic resonance rather than genes.

Memes and morphic fields

In the standard materialist view, all inheritance is material except cultural inheritance, which everyone agrees works in a different way, primarily through animals and humans learning by imitation. In 1976, Richard Dawkins proposed the term 'meme' for a unit of cultural inheritance, by analogy with gene:

Examples of memes are tunes, ideas, catch-phrases, clothes fash-
ions, ways of making pots or of building arches. Just as genes
propagate themselves in the gene pool by leaping from body to
body via sperms or eggs, so memes propagate themselves in the
meme pool by leaping from brain to brain via a process which, in
the broad sense, can be called imitation.[46]

This idea has itself proved to be a successful meme, showing that
there is a need for such a concept.[47] The materialist philosopher
Daniel Dennett has used the meme concept as the 'cornerstone' of
his theory of mind.[48] But the term meme is too atomistic and
reductionistic, and various authors have proposed new terms to
refer to complexes of memes linked together in larger structures,
like 'co-adapted meme complex' or 'memeplex'.[49]

Atheists are particularly keen on the idea of religions as meme
complexes, and think of them as like viruses infecting other people's
brains.[50] They regard themselves as immune. But materialism must
itself be a virus-like meme complex that infects materialists' brains.
When the materialist memeplex is particularly virulent, it turns its
victims into proselytising atheists so that it can jump from their
brains to as many other people's brains as possible.

Despite all the speculation about memes and their role in culture
and religion, their nature has remained obscure. Materialists like to
think of them as material structures inside material brains, but no
one has ever found a meme inside a brain, or seen one leaping from
one brain to another. They are invisible. They are in fact patterns of
organisation or information, and I propose that it is more helpful to
think of them as morphic fields, transferred from brain to brain by
morphic resonance.[51] From the materialist point of view there is a
fundamental difference of kind between genetic inheritance and
cultural inheritance, because the former is material and the latter is
not. Thinking of memes as if they are material objects is an attempt
to overcome this problem, but it is a rhetorical manoeuvre, not a
scientifically testable hypothesis.

I once attempted to discuss this point with Richard Dawkins. I
said to him that memes and morphic fields seemed to play a

similar role in cultural inheritance. He replied, 'They have nothing in common whatsoever. Memes are real because they are material. They exist inside material brains. Morphic fields are not material and therefore they don't exist.'[52] That was the end of the discussion. But like morphic fields, memes could only work through patterns of brain activity. Memes cannot possibly be material objects, like little computer chips or miniature CDs.

From the point of view of morphic resonance, there is only a difference of degree, not of kind, between the hereditary transmission of form and behaviour and the cultural transmission of patterns of behaviour. Both depend on morphic resonance. Morphic fields are not atomistic and particulate, but organised in nested hierarchies or holarchies, which fit much more naturally with the structure of culturally inherited patterns. Language, for example, is made up of a nested hierarchy of levels: phonemes, in words, in syllables, in phrases, in sentences (Figure 1.1). I return to a discussion of the role of morphic fields in minds and memories in Chapter 7.

What difference does it make?

The belief that genes are the basis of almost all inheritance is not just an intellectual theory but has had enormous economic and political consequences. It has resulted in the investment of hundreds of billions of dollars in genome and biotechnology projects. If genes are the keys to life, then people want to own them and exploit them. But if genes are grossly overrated, genomics will never live up to the high hopes it once engendered. A few companies make useful products, but many make promises that never come true.

The gene-centred view of life has dominated science since the 1960s with baleful effects on our general culture. Jeffrey Skilling, the CEO of Enron, a corporation noted for its greed and predatory behaviour, said that his favourite book was *The Selfish Gene*,[53] and the selfish-gene theory was a major part of Enron's corporate culture until the company collapsed in 2001. Skilling, who is serving a long jail sentence, interpreted neo-Darwinism to mean that selfishness was ultimately good even for its victims, because it weeded out losers and forced survivors to become strong.[54]

Genes are not individualistic and selfish, despite the rhetoric suggesting they are. As parts of larger wholes, they work co-operatively in the development and functioning of organisms. If they have any moral message for humans, it is that life depends on working together and not on ruthless competition.

A broader understanding of heredity that includes genes, gene modifications and morphic resonance opens up many new questions, and helps free the life sciences from the tunnel vision of molecular biology. It makes a big difference scientifically. For a start, the word 'hereditary' is no longer synonymous with 'genetic'. Genes are part of heredity, not all of it. Morphic resonance may underlie the inheritance of form, and behaviour. This resonance is physical, but not material. Similarly, morphic resonance may play a major role in cultural inheritance.

Through morphic resonance, animals and plants are connected with their predecessors. Each individual both draws upon and contributes to the collective memory of the species. Animals and plants inherit the habits of their species and of their breed. The same applies to humans.

An extended understanding of heredity changes the way we think of ourselves, the influence of our predecessors, and our effects on generations yet unborn.

Questions for materialists

Do you agree with Lewis Wolpert that 'By 1 May 2029, given the genome of a fertilised egg of an animal or plant, we will be able to predict in at least one case all the details of the organism that develops from it, including any abnormalities'? If so, how much would you be prepared to bet on it?

If you believe genes 'program' organisms, how do you think the programs work?

Do you think that mathematical models will eventually explain the inheritance of form and behaviour? If so, are organisms 'reifications' of mathematics?

How do you think the missing heritability problem might be solved?

SUMMARY

Genes are overrated in the sense that they do not 'code for' or 'program' the form and behaviour of organisms. They specify the sequence of amino acids in protein molecules and some are involved in the control of protein synthesis. The Human Genome Project and other genome projects have been disappointing, both scientifically and financially, because they were based on a false conception of what genes do. The inheritance of development and behaviour may depend on organising fields that have an inherent memory. In addition, characteristics acquired by plants and animals can be passed on to their descendants epigenetically, through modifications of gene expression rather than mutation. Habits of growth and behaviour can be inherited through a collective memory of the species, on which each individual draws, and to which it contributes: organisms inherit habits of form and behaviour that are not coded in the genes by the process of morphic resonance. Morphic resonance may also underlie cultural inheritance, which differs in degree, but not in kind, from the inheritance of forms and instincts.

7

Are Memories Stored as Material Traces?

We take memory for granted, like the air we breathe. Everything we do, see and think is shaped by habits and memories. My ability to write this book, and yours to read it, presupposes the memory of words and their meanings. My ability to ride a bicycle depends on unconscious habit memory. I can recall facts I have learned, like the year of the Battle of Hastings – 1066; I can recognise people I first met years ago; I can remember specific incidents that happened when I was on holiday in Canada last summer. These are different kinds of memory, but all involve influences from the past that affect me in the present. Our memories underlie all our experience. And obviously animals have memories too.

How does memory work? Most people take it for granted that memories must somehow be stored in brains as material traces. In ancient Greece these traces were usually compared to impressions in wax. In the early twentieth century they were compared to connections between wires in a telephone exchange, and now they are thought of by analogy with memory storage systems in computers. Although the metaphors change, the trace theory is taken for granted by most scientists, and almost everyone else.

From a materialist point of view, memories *must* be stored as material traces in brains. Where else could they be? The neuroscientist Steven Rose expressed the standard assumptions as follows:

Memories are in some way 'in' the mind, and therefore, for a biologist, also 'in' the brain. But how? The term memory must include at least two separate processes. It must involve, on the one hand, that of learning something new about the world around us; and on the other, at some later date, *recalling*, or remembering that thing. We infer that what lies between the learning and the remembering must be some permanent record, a *memory trace*, within the brain.[1]

This seems obvious and straightforward. It might seem pointless to question it. Yet the trace theory of memory is very questionable indeed. It raises appalling logical problems. Attempts to locate memory traces have been unsuccessful despite more than a century of research, costing many billions of dollars. For promissory materialists, this failure does not imply that the trace theory of memory might be wrong; it merely means that we need to spend more time and money searching for the elusive memory traces.

But memory traces are not the only option. Several philosophers in the ancient world, notably Plotinus, were sceptical that memories were material impressions,[2] and argued that they were immaterial rather than material, aspects of the soul rather than the body.[3] Likewise, more recent philosophers, like Henri Bergson and Alfred North Whitehead, saw memories as direct connections across time, not material structures in brains (see Chapter 4).

My own suggestion is that memories depend on morphic resonance. All individuals are influenced by morphic resonance from their own past. Morphic resonance depends on similarity; since organisms are more similar to themselves in the past than to other members of their species, self-resonance is highly specific. Individual memory and collective memory both depend on morphic resonance; they differ from each other in degree, not in kind.

I start with the trace theory of memory, then discuss the resonance hypothesis, and finally ways in which this hypothesis can be tested.

Logical and chemical problems

Several modern philosophers have pointed out that the trace theory of memory runs into an insoluble logical problem, quite apart from repeated failures to find memory traces.

In order for a memory trace to be consulted or reactivated, there has to be a retrieval system, and this system needs to identify the stored memory it is looking for. To do so it must recognise it, which means the retrieval system must itself have a memory. There is therefore a vicious regress: if the retrieval system is endowed with a memory store, this in turn requires a retrieval system with memory, and so on *ad infinitum*.[4]

There is a structural problem too. Memories can persist for decades, yet the nervous system is dynamic, continually changing, and so are the molecules within it. As Francis Crick put it, 'Almost all the molecules in our bodies, with the exception of DNA, the genetic material, turn over in a matter of days, weeks, or at the most a few months. How then is memory stored in the brain so that its trace is relatively immune to molecular turnover?' He suggested a complex mechanism whereby molecules were replaced one at a time so as to preserve the overall state of the memory-storage structures.[5] No such mechanism has been detected.

For decades, the most popular theory has been that memory must depend on changes in connections between nerve cells, the synapses. Yet attempts to locate memory stores have proved unsuccessful over and over again.

The fruitless search for memory traces

In the 1890s, Ivan Pavlov studied the way that animals such as dogs could learn to associate a stimulus, such as hearing a bell, with being fed. After repeated training, merely hearing the bell could cause the dogs to salivate. Pavlov called this a conditioned reflex. For many scientists at the time, this research suggested that the animals' memory depended on reflex arcs, in which the nerve fibres were like wires and the brain like a telephone exchange. But

Pavlov himself was reluctant to claim there were specific localised traces. He discovered that conditioning could survive massive surgical damage to the brain.[6] Those who knew less about it were less cautious, and in the first few decades of the twentieth century many biologists assumed that *all* psychological activity, including the phenomena of the human mind, could ultimately be reduced to chains of reflexes wired together in the brain.

In a heroic series of experiments lasting more than thirty years, Karl Lashley (1890–1958) tried to locate specific memory traces, or 'engrams', in the brains of rats, monkeys and chimpanzees. He trained the animals in a variety of tasks ranging from simple conditioned reflexes to the solution of difficult problems. After the training, he surgically cut nerve tracts or removed portions of the brain and measured the effects on the animals' memory. To his astonishment, he found that the animals could still remember what they had learned even after large amounts of brain tissue had been removed.

Lashley first became sceptical of the supposed path of conditioned reflex arcs through the motor cortex when he found that rats trained to respond in specific ways to light could perform almost as well as control rats after almost all their motor cortex was cut out. In similar experiments with monkeys, he removed most of the motor cortex after they had been trained to open boxes with latches. This operation resulted in a temporary paralysis. After two or three months, when they recovered their ability to move in a co-ordinated way, they were exposed to the puzzle boxes again. They opened them promptly without random exploratory movements.

Lashley then showed that learned habits were retained after the associative areas of the brain were destroyed. Habits also survived a series of deep incisions into the cerebral cortex that destroyed cross-connections within it. Moreover, if the cerebral cortex was intact, removal of subcortical structures such as the cerebellum did not destroy the memory either.

Lashley started as an enthusiastic supporter of the reflex theory of learning, but was forced to abandon it:

The original programme of research looked towards the tracing of conditioned-reflex arcs throughout the cortex . . . The experimental findings have never fitted into such a scheme. Rather, they have emphasised the unitary character of every habit, the impossibility of stating any learning as concatenations of reflexes, and the participation of large masses of nervous tissue in the functions rather than the development of restricted conduction paths.[7]

Lashley suggested that

the characteristics of the nervous network are such that when it is subject to any pattern of excitation, it may develop a pattern of activity, reduplicated through an entire functional area by spread of excitations, such as the surface of a liquid develops an interference pattern of spreading waves when it is disturbed at several points.

He suggested that recall involved 'some sort of resonance among a very large number of neurons'.[8] These ideas were carried further by his former student Karl Pribram in his proposal that memories are stored in a distributed manner throughout the brain analogous to the interference patterns in a hologram.[9]

Even in invertebrates specific memory traces have proved elusive. In a series of experiments with trained octopuses, learned habits survived when various parts of the brain were removed, leading to the seemingly paradoxical conclusion that 'memory is both everywhere and nowhere in particular'.[10]

Despite these results, new generations of researchers have tried again and again to find localised memories. In the 1980s, Steven Rose and his colleagues thought they had at last succeeded in finding traces in the brains of day-old chicks. They trained the chicks to avoid pecking at little coloured lights by making them sick, and the chicks duly avoided these stimuli when they encountered them again. Rose and his colleagues then studied the changes in the brains of these chicks, and found that nerve cells in a particular region of the left forebrain underwent more active growth and development when learning took place than when it did not.[11]

These findings agreed with results from studies of the growing brains of young rats, kittens and monkeys, which found that active nerve cells in the brain developed more than inactive nerve cells. But the greater development of active cells did not prove that they contained specific memory traces. When the region of active cells was surgically removed from the chicks' left forebrains a day after training, the chicks could still remember what they had learned. Therefore the region of the brain involved in the learning process was not necessary for the retention of memory. Once again, the hypothetical memory traces proved elusive, and once more those who searched for them were forced to postulate unidentified 'storage systems' somewhere else in the brain.[12]

In a more recent series of studies, mice were studied as they learned to negotiate a maze. The formation of memories involved activity in the median temporal lobes of the brain, particularly in the hippocampus. The ability to form long-term memories depended on a process called long-term potentiation, which involved protein synthesis in hippocampal nerve cells. But yet again, the memories proved elusive. Once the memories had been established, the destruction of the hippocampus on both sides of the brain failed to wipe them out. Thus, the researchers concluded, the hypothetical memory traces must somehow have moved from one part of the brain to another.

Erik Kandel, who won the Nobel Prize in 2000 for his work on memory in the sea slug, *Aplysia*, drew attention to some of these problems in his acceptance speech:

How do different regions of the hippocampus and the median temporal lobe . . . interact in the storage of explicit memory? We do not, for example, understand why the initial storage of memory requires the hippocampus, whereas the hippocampus is not required once a memory has been stored for weeks or months. What critical information does the hippocampus convey to the neo-cortex? We also know very little about the recall of explicit (declarative) memory . . . These systems properties of the brain will require more than the bottom-up approach of molecular biology.[13]

Currently, in the Connectome Project researchers at the Massachusetts Institute of Technology and elsewhere are trying to map some of the trillions of connections between nerve cells in mammalian brains, using thin slices of brain tissue and sophisticated computer analyses of the images. There are about 100 billion neurons in the human brain. As Sebastian Seung, the leader of the MIT team, pointed out, 'In the cerebral cortex, it's believed that one neuron is connected to 10,000 others.' This is a vastly ambitious project, but it seems unlikely to shed light on memory storage. First of all, a person has to be dead before his brain can be cut up, so changes before and after learning cannot be studied in this way. Second, there are great differences between the brains of different people; we do not have identical 'wiring'.

The same is true of small animals like mice. A pilot project in the Max Planck Institute in Germany looked at the wiring diagrams for just fifteen neurons that control two small muscles in mouse ears. Even though this work was a technical *tour de force*, it revealed no unique wiring diagram. Even for the right and left ears of the same animal the patterns of connection were different.[14]

The most striking deviations from normal brain structure occur in people who suffered from hydrocephalus when they were babies. In this condition, also called 'water on the brain', much of the skull is filled with cerebrospinal fluid. The British neurologist John Lorber found that some people with extreme hydrocephalus were surprisingly normal, which led him to ask the provocative question: 'Is the brain really necessary?' He scanned the brains of more than six hundred people with hydrocephalus, and found that about sixty had more than 95 per cent of the cranial cavity filled with cerebrospinal fluid. Some were seriously retarded, but others were more or less normal, and some had IQs of well over

100. One young man who had an IQ of 126 and a first-class degree in mathematics, a student from Sheffield University, had 'virtually no brain'. His skull was lined with a thin layer of brain cells about a millimetre thick, and the rest of the space was filled with fluid.[15] Any attempt to explain his brain in terms of a standard 'connectome' would be doomed to failure. His mental activity and his memory were still able to function more or less normally even though he had a brain only five per cent of the normal size.

The available evidence shows that memories cannot be explained in terms of localised changes in synapses. Brain activity involves rhythmic patterns of electrical activity extended over thousands or millions of nerve cells, rather than simple reflex arcs like wires in a telephone exchange or wiring diagrams of computers. These patterns of nervous activity set up – and respond to – changes in the electromagnetic fields in the brain.[16] The oscillating fields of entire brains are routinely measured in hospitals with electroencephalographs (EEG), and within these overall rhythms there are many subsidiary patterns of electrical activity in different regions of the brain. If these patterns, or systems properties, are to be remembered, resonance across time seems more likely than chemical storage in nerve endings.

More than a century of intensive, well-funded research has failed to pin down memory traces in brains. There may be a very simple reason for this: the hypothetical traces do not exist. However long or hard researchers look for them they may never find them. Instead, memories may depend on morphic resonance from an organism's own past. The brain may be more like a television set than a hard-drive recorder. What you see on TV depends on the resonant tuning of the set to invisible fields. No one can find out today what programmes you watched yesterday by analysing the wires and transistors in your TV set for traces of yesterday's programmes.

For the same reason, the fact that injury and brain degeneration, as in Alzheimer's disease, lead to loss of memory does not prove that memories are stored in the damaged tissue. If I snipped a wire or removed some components from the sound circuits of your TV set, I could render it speechless, or aphasic. But this would not mean that all the sounds were stored in the damaged components.

Can a moth remember what it learned as a caterpillar?

Insects that undergo complete metamorphosis experience enormous changes in anatomy and lifestyle. It is hard to believe that a caterpillar chewing a leaf is the same organism as the moth that later emerges from the pupa. In the pupa, almost all the caterpillar tissues are dissolved before the new structures of the adult develop. Most of the nervous system is dissolved as well.

In a recent study, Martha Weiss and her colleagues at Georgetown University, Washington, found that moths could remember what they had learned as caterpillars in spite of all the changes they went through during metamorphosis. They trained caterpillars of the Carolina Sphinx moth, *Manduca sexta*, to avoid the odour of ethyl acetate by associating exposure to this odour with a mild electric shock. After two larval moults and metamorphosis within the pupae, the adult moths were averse to ethyl acetate, despite that radical transformation of their nervous system. Weiss and her colleagues carried out careful controls that showed this was a real transfer of learning, not just a carryover of odours absorbed by the tested caterpillars.[17]

This ability of adult moths to remember their experience as caterpillars may well be of evolutionary significance. If the plants that moths have experienced as caterpillars influence the behaviour of adults, the female moths will tend to avoid laying eggs on harmful plants and favour nutritious ones, even if members of the species have never encountered these plants before. New patterns of preference for particular host plants could be established in a single generation, and would persist in their offspring; a species could evolve new feeding habits very rapidly.

The carryover of learning from caterpillar to moth after the dissolution of most of the nervous system would be very puzzling indeed if all memories were stored as material traces, but there is already evidence from higher animals and humans that memories may not be stored in traces and can survive substantial damage to brains.

Brain damage and loss of memory

Brain damage can result in two kinds of memory loss: retrograde (backwards) amnesia, forgetting what happened before the damage, and anterograde (forwards) amnesia, losing the ability to remember what happens after the damage.

The best-known examples of retrograde amnesia occur after concussion. As a result of a sudden blow on the head a person loses consciousness and becomes paralysed for a few seconds or for many days, depending on the severity of the impact. As she recovers and regains the ability to speak, she may seem normal in most respects, but is unable to recall what happened before the accident. Typically, as recovery proceeds, the first of the forgotten events to be recalled are those longest ago; the memory of more recent events returns progressively.

In such cases, amnesia cannot be due to the destruction of memory traces, for the lost memories return. Karl Lashley reached a similar conclusion years ago:

> I believe that the evidence strongly favours the view that amnesia from brain injury rarely, if ever, is due to the destruction of specific memory traces. Rather, the amnesias represent a lowered level of vigilance, a greater difficulty in activating the organised pattern of traces, or a disturbance of some broader system of organised functions.[18]

Although many memories return, the events immediately preceding a blow on the head may never be recovered: there may be a permanent blank period. For example, a motorist may remember approaching the crossroads where an accident occurred, but nothing more. A similar 'momentary retrograde amnesia' also occurs as a result of electroconvulsive therapy, administered to some psychiatric patients by passing a burst of electric current through their heads. They usually cannot remember what happened immediately before the administration of the shock.[19]

Events and information in short-term memory are forgotten because a loss of consciousness prevents them being connected up into patterns of relationship that can be remembered. The failure to make such connections, and hence to turn short-term memories into long-term memories, often persists for some time after a concussed patient has regained consciousness, and is sometimes described as 'memorising defect'. People in this condition rapidly forget events almost as soon as they occur.

Everyone agrees that the formation of memories is an active process. Either the inability to construct them prevents new memory traces being formed; or this inability prevents the formation of new morphic fields, resonant patterns of activity, and if these patterns are not formed in the first place, they cannot be recalled by morphic resonance.

Some kinds of brain damage have very specific effects on people's abilities to recognise and recall,[20] and others cause specific disorders, such as aphasias (disorders of language use) resulting from lesions in various parts of the cortex in the left hemisphere. These kinds of damage disturb the organised patterns of activity in the brain,[21] and affect the brain's ability to tune in to skills and memories by morphic resonance.

Holograms and the implicate order

In a famous series of investigations carried out during brain surgery on conscious patients, Wilder Penfield and his colleagues tested the effects of mild electrical stimulation of various regions of the cerebral cortex. As the electrode touched parts of the motor cortex, limbs moved. Electrically stimulating the auditory or visual cortex evoked auditory or visual hallucinations like buzzing noises or flashes of light. Stimulation of the secondary visual cortex gave hallucinations of flowers, for example, or animals, or familiar people. When some regions of the temporal cortex were stimulated, some patients recalled dream-like memories, for example of a concert or a telephone conversation.[22]

Penfield initially assumed that the electrical evocation of memories meant that they were stored in the stimulated tissue, which he named the 'memory cortex'. On further consideration, he changed his mind: 'This was a mistake . . . The record is not in the cortex.'[23] Like Lashley and Pribram, he gave up the idea of localised memory traces in favour of the theory that they were widely distributed in other parts of the brain.

The most popular analogy for distributed memory storage is holography, a form of lens-less photography in which interference patterns are stored as holograms, from which the original image can be reconstructed in three dimensions. If part of the hologram is destroyed, the whole image can still be reconstructed from the remaining parts, although in lower definition. The whole is present in each part. This may sound mysterious, but the basic principle is simple and familiar. As you look around you now, your eyes are sampling light from all the parts of the scene in front of you. The light absorbed by your eyes is only a small part of the available light, and yet you can see the whole scene. If you move a few feet, you can still see everything, the whole scene is present there too, although you are now sampling the light-waves in a different place. In a similar way, the whole is enfolded into each part of a hologram. This is not true of an ordinary photograph: if you tear off half the photo, you have lost half the image. If you tear off half a hologram, the whole image can still be re-created.

But what if the holographic wave-patterns are not stored in the brain at all? Pribram later came to this conclusion, and thought of the brain as a 'wave-form analyser' rather than a storage system, comparing it to a radio receiver that picked up wave-forms from the 'implicate order', rendering them explicate.[24] This aspect of his thinking was influenced by the quantum physicist David Bohm, who suggested that the entire universe is holographic, in the sense that wholeness is enfolded into every part.[25]

According to Bohm, the observable or manifest world is the explicate or unfolded order, which emerges from the implicate or enfolded order.[26] Bohm thought that the implicate order contains a kind of memory. What happens in one place is 'introjected' or 'injected' into the implicate order, which is potentially present everywhere; thereafter when the implicate order unfolds into the explicate order, this memory affects what happens, giving the process very similar properties to morphic resonance. In Bohm's words, each moment will 'contain a projection of the re-injection of the previous moments, which is a kind of memory; so that would result in a general replication of past forms'.[27]

Maybe morphic resonance will one day be included in an enlarged version of quantum theory, as Bohm suggested. No one yet knows. The question 'How can morphic resonance be explained?' is open. In the context of a debate about the reality of memory traces, does morphic resonance – or memory in the implicate order – fit the facts better than the trace theory?

Resonance across time

The trace theory says that memories are stored materially in brains, for example as chemicals in synapses. The alternative is the resonance theory: memories are transferred by resonance from similar patterns of activity in the past. We tune in to ourselves in the past; we do not carry our memories around inside our heads.

The resonance of memory is part of a much wider hypothesis. The hypothesis of morphic resonance proposes a resonance across space and time of patterns of vibratory activity in all self-organising systems.[28] Morphic resonance underlies habits of crystallisation and protein folding (see Chapter 3). It also underlies the inheritance of morphogenetic fields and of patterns of instinctive behaviour (see Chapter 6). It plays an essential role in the transfer of learning, as discussed below. Morphic resonance provides a new way of looking at memories. There are at least five kinds of memory: habituation, sensitisation, behavioural memory, recognition and recalling.

Habituation and sensitisation

Habituation means getting used to things. If you hear a new sound, or smell a new smell you may pay attention to it to start with, but if it makes no difference, you soon cease to notice it. You don't notice the pressure of your clothes on your body most of the time, or the pressure of your bottom on the seat on which you are sitting, or the sounds of a clock ticking, or the many other background noises around you.

Habituation is one of the most fundamental kinds of memory and underlies all our responses to our environment. Generally speaking, we do not notice what stays the same; we notice changes or differences. All our senses work on this principle. If you are gazing over a landscape, anything that moves immediately catches your eye. If there is a change in the background noise, you notice it. Our entire culture works on the same principle, which is why gossip and newspapers rarely concern themselves with things that stay the same. They are about changes or differences.

Other animals likewise become accustomed to their environments. They generally react to something new because they are not used to it, often showing alarm or avoidance. This kind of response even occurs in single-celled animals like *Stentor*, which lives in marshy pools. Each *Stentor* is a trumpet-shaped cell covered with rows of fine, beating hairs called cilia. The activity of the cilia sets up currents around the cell, carrying suspended particles to the mouth, which is at the bottom of a tiny vortex (Figure 7.1). These cells are attached at their base by a 'foot', and the lower part of the cell is surrounded by a mucus-like tube. If the surface to which it is attached is slightly jolted, *Stentor* rapidly contracts into its tube. If nothing happens, after about half a minute it extends again and the cilia resume their activity. If the same stimulus is repeated, it does not contract but continues its normal activities. This is not a result of fatigue because the cell responds to a new stimulus, such as being touched, by contracting again.[29]

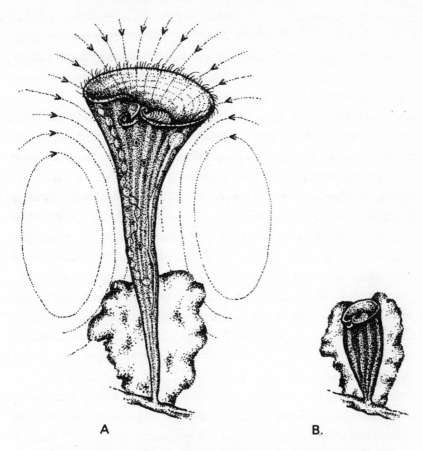

A   B.

FIGURE 7.1A: The single-celled organism *Stentor raesilii*, showing the currents of water around it caused by the beating of its cilia. In response to an unfamiliar stimulus it rapidly contracts into its tube (B). (After Jennings, 1906)

The cell membranes of *Stentor* have an electrical charge across them, just like nerve cells. When they are stimulated, an action potential sweeps over the surface of the cell, very similar to a nerve impulse, and this leads to the cell contracting.[30] As it becomes habituated, the receptors on the cell's membrane become less sensitive to mechanical stimulation, and the action potential is not triggered.[31] Since *Stentor* is a single cell, its memory cannot be explained in terms of changes in nerve endings, or synapses, because it has none.

Habituation implies a kind of memory that enables harmless and irrelevant stimuli to be recognised when they recur. Morphic resonance suggests a straightforward explanation. The organism is in resonance with its own past patterns of activity, including its return to normal following its withdrawal response to a harmless stimulus. When the stimulus is repeated, the organism resonates with its previous pattern of response, including the return to normal activity. It returns to normal activity sooner, and responds less and less, until the harmless stimuli are ignored. It habituates through self-resonance. A new stimulus stands out precisely because it is new and unfamiliar.

Habituation occurs in all animals, large and small, with and without nervous systems. The effects of habituation have been studied in detail in the giant marine slug *Aplysia*, which grows more than a foot long. Its nervous system is relatively simple, and is similar in different individuals. Normally the slug's gill is extended, but if the slug is touched, the gill is withdrawn. This reflex soon ceases if harmless stimuli are repeated; the slugs habituate, just like *Stentor*. Eric Kandel and his group showed that only four motor nerve cells are responsible for the gill withdrawal response. As habituation occurs, the sensory nerve cells cease to excite the motor cells because they release fewer and fewer packets of chemical transmitter at the synapses with the motor cells. But the fact that the synapses function differently as a result of habituation does not prove that the memory is stored chemically in the synapses. The entire system may habituate as a result of self-resonance, as in *Stentor*. Self-resonance may underlie habituation in animals at all levels of complexity, including ourselves.

Sensitisation is the opposite of habituation: animals become more responsive to stimuli that have a harmful effect. Again, even single-celled animals like *Stentor* exhibit this kind of behaviour. If a stream of noxious particles is directed at *Stentor*, it contracts into its tube. The next time it is exposed to the same particles it contracts more rapidly, and after several exposures, it goes on contracting inside its tube until its foot is detached; it swims away until it finds a more peaceful place to settle down, where it builds a new tube and resumes its normal life. *Aplysia* shows a similar kind of sensitisation, and Kandel and his group have described several changes that occur in the nerve cells as this happens. Whereas habituation results in less neurotransmitter being released by sensory neurons in their synapses with motor neurons, sensitisation results in more being released.[32]

Again, there is no need to suppose that the memory that underlies sensitisation is stored in the form of chemical changes inside the cells. Like habituation, sensitisation fits well with a self-resonance model. When a stimulus proved harmful in the past occurs, the organism resonates with itself, responding to the same stimulus, resulting in a greater response. In addition, sensitisation can reach a threshold where the organism does something different. *Stentor* swims away.[33] *Aplysia* releases toxic ink containing hydrogen peroxide.[34]

Resonant learning

Many animals learn patterns of behaviour from other members of their group through imitation. For example, some species of bird, like blackbirds, learn parts of songs by listening to the songs of nearby adults. This is a kind of cultural inheritance.

Cultural inheritance reaches its highest development in humanity where all human beings learn a great variety of patterns of behaviour, including the use of language, as well as many physical and mental skills, like doing arithmetic, playing the flute or knitting. From the point of view of morphic resonance, the transfer of these skills is a kind of resonance process.

In the 1980s, neuroscientists discovered that when animals watched other animals carrying out a particular action, changes in the motor part of their brains mirrored those in the brains of the animals they were watching. These responses are often described in terms of 'mirror neurons': the brain activity mirrors that of the animal being watched, and involves the same sorts of changes that take place in carrying out the action itself. But the term mirror neuron is misleading if it suggests that special kinds of nerves are required for this activity. Instead, it is better thought of as a kind of resonance. In fact, Vittorio Gallese, one of the discoverers of mirror neurons, refers to the imitation of movements or actions by another individual as 'resonance behaviour'.[35]

Resonance behaviour is a new phrase, but the phenomenon itself is not a new discovery. The entire pornography industry depends on it. Watching other people engaged in sexual activity stimulates erotic arousal by a kind of resonance.

Some neuroscientists have extended the idea of mirror systems to what they call a 'motor resonance theory of mind reading', whereby the nervous system responds 'to execution and observation of goal-oriented actions'.[36] This resonance is not confined to the brain but to the entire pattern of movements of the body as well, and no doubt plays a major part in the learning of skills, such as riding a bicycle, and in other forms of 'learning by doing'.

Through repetition, behavioural patterns and skills improve, and become increasingly habitual. Both the acquisition of new patterns of behaviour and remembering them fit well with a resonance model.

Recognising

Recognition involves the awareness that a present experience is also remembered: we *know* that we were in this place before, or met this person somewhere, or came across this fact or idea. But we may not be able to recall where or when, or recall the person's or the place's name. Recognition and recall are different kinds of memory: recognition depends on a similarity between present experience and previous experience. Recall involves an active reconstruction of the past on the basis of remembered meanings or connections.

Recognising is easier than recalling. For example, it is usually easier to recognise people than remember their names. Most of us have remarkable powers of recognition that we usually take for granted. Many laboratory experiments have demonstrated just how powerful this ability can be. For example, in one study, subjects were asked to memorise a meaningless shape. When they were asked to recall it by drawing it, their ability to do so declined rapidly within minutes. By contrast, most people could pick out the test shape from a range of similar shapes weeks later.[37]

Recognition, like habituation, depends on morphic resonance with previous similar patterns of activity. The pattern of vibratory activity within your sensory organs and nervous system when you see a person you know is similar to the pattern when you saw the same person before. The sensory stimuli are similar and have similar effects on the sense organs and the nervous system. The greater the similarity, the stronger the resonance.

Recalling

Conscious recall is an active process. The ability to recall a particular experience depends on the ways we made connections in the first place. To the extent that we use language to categorise and connect the elements of experience, we can use language to help reconstruct these past patterns. But we cannot recall connections that were not made to start with.

Our short-term memory for words and phrases enables us to remember them long enough to grasp their connections and understand their meanings. We usually remember meanings – patterns of connection – rather than the actual words. It is relatively easy to summarise the gist of a recent conversation but, for most of us, impossible to reproduce it verbatim. The same is true of written language: you may recall some of the facts and ideas in the preceding chapters of this book, but you will probably recall very few passages word for word.

Short-term memories provide the opportunities for elements of our near-present experience to be connected with each other, as well as with past experience. What is not connected is forgotten. Short-term memory is often compared to a computer's RAM (Random Access Memory), and has a very limited capacity, typically 7±2 items. In the 1940s, the neuroscientist Donald Hebb pointed out that such short-term memories, lasting less than a minute, were unlikely to be stored chemically and suggested that they might depend on reverberating circuits of electrical activity – again implying a process of resonance.

In the case of spatial recall – for instance, in remembering the layout of a particular house – the connections between different spaces are related to movements of the body; for example, along a corridor, climbing stairs and entering a room.

The principles of memorising and recalling have long been understood; the basic principles of mnemonic systems were well known in classical times and were taught to students of rhetoric, providing techniques for establishing connections that enable items to be recalled more easily.[38] Some methods depend on verbal connections and involve coding the information in rhymes, phrases or sentences. For instance, 'Richard Of York Gained Battles In Vain' is a well-known mnemonic for the colours of the rainbow (Red, Orange, Yellow, Green, Blue, Indigo, Violet). Other systems are spatial and rely on visual imagery. For instance, in the 'method of loci' one first memorises a sequence of locations; for example, the various rooms and cupboards of one's own house. Each item to be recalled is then visualised in one of these locations, and remembered by imagining walking from one place to the other and finding the object there. Modern mnemonic systems, such as systems for improving your memory power advertised in popular magazines, are the heirs of this long and rich tradition.[39]

Memorising spatial patterns in many animals depends on the activity of the hippocampus, as discussed above, and the activity of the brain in this and other regions seems to be necessary for connecting together the items to be recalled. Between being laid down and recalled, the memories are usually supposed to be encoded in elusive long-term memory traces. The resonance hypothesis fits the facts better. The pattern of connections established when the memories are formed is associated with rhythmic patterns of brain activity. The memories are recalled through similar patterns of activity established by morphic resonance. They are not stored as traces in the brain.

Experimental tests

If memories are stored in individual animals' brains, then anything an animal learns is confined to its own brain. When it dies the memory is extinguished. But if memory is a resonant phenomenon through which organisms specifically resonate with themselves in the past, individual memory and collective memory are different aspects of the same phenomenon; they differ in degree not in kind.

This hypothesis is testable. If rats learn a new trick in one place, then rats all over the world should be able to learn the same trick quicker. The more rats that learn it, the easier it should become everywhere else. There is already evidence from one of the longest series of experiments in the history of psychology that rats do indeed seem to learn quicker what other rats have already learned. The more that learned to escape from a water maze, the easier it became for others to do so. These experiments, conducted first at Harvard, then at Edinburgh and Melbourne universities, showed that the Scottish and Australian rats took up more or less where the Harvard rats had left off, and their descendants learned even faster. Some got it right first time with no need for learning at all. In the experiment at Melbourne University, a line of control rats, whose parents had never been trained, showed the same pattern of improvement as rats descended from trained parents, showing that this effect was not passed through the genes, or through epigenetic modifications of genes. All similar rats learned quicker, just as the hypothesis of morphic resonance would predict.[40]

Likewise, humans should be able to learn more easily what others have already learned. New skills like snowboarding and playing computer games should become easier to learn, on average. Of course there will always be faster and slower learners, but the general tendency should be towards quicker learning. Much anecdotal evidence suggests that this is so. But for hard, quantitative evidence, the best place to look is in standardised tests that have remained more or less the same over decades. Intelligence quotient (IQ) tests are a good example. By morphic resonance, the questions should become easier to answer because so many people have answered them before. The scores in the tests should rise not because people are becoming more intelligent but because the tests are becoming easier to do. Just such an effect has in fact occurred and is known as the Flynn effect after the psychologist James Flynn who has done so much to document this phenomenon.[41] Average IQ test scores have been rising for decades by 30 per cent or more. Data from the United States are in Figure 7.2.

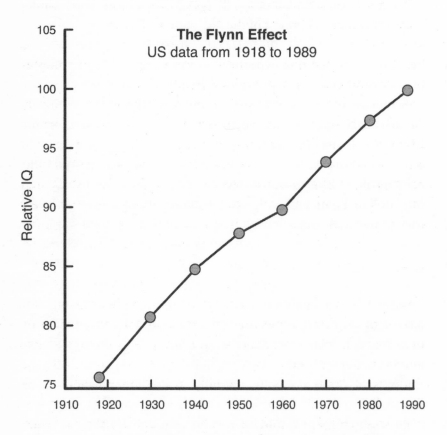

FIGURE 7.2. THE FLYNN EFFECT: changes in average IQ scores in the United States, relative to 1989 values.[42]

There has been a long debate among psychologists about possible reasons for the Flynn effect. Attempted explanations in terms of nutrition, urbanisation, exposure to TV and practice with examinations seem to account for only a small part of this effect. At first Flynn confessed himself baffled, and has tried out a number of ever more complex explanations. His most recent attempt ascribes this effect to a change in the general culture:

> The best short-hand description I can offer is this. During the twentieth century, people invested their intelligence in the solution of new cognitive problems. Formal education played a proximate causal role but a full appreciation of causes involves grasping the total impact of the industrial revolution.[43]

The trouble is that this hypothesis is vague, obscure and untestable. Morphic resonance provides a simpler explanation.

Scientists in universities in Europe and America have already carried out a series of tests specifically designed to test for morphic resonance in human learning, particularly in connection with written languages. Most have given positive, statistically significant results.[44] This is inevitably a controversial area of research but, unlike Flynn's hypothesis, morphic resonance is relatively easy to test with animals and people.

What difference does it make?

I find it makes a big difference to think of tuning in to my memories, instead of retrieving them from stores inside my brain by obscure molecular mechanisms. Resonance feels more plausible and fits better with experience. It is also more compatible with the findings of brain research: memory traces are nowhere to be found.

In research there would be a shift of focus from the molecular details of nerve cells to the transfer of memory by resonance. This shift would also open up the question of collective memory, which the psychologist C. G. Jung thought of in terms of the collective unconscious.

If learning involves a process of resonance not only with the teacher who is transmitting the skill, but all those who have learned it before, educational methods could be improved by deliberately enhancing the process of resonance, leading to a faster and more effective transfer of skills.

The resonance theory of memory also opens up a religious question. All religions take it for granted that some aspect of a person's memory survives that person's bodily death. In Hindu and Buddhist theories of reincarnation or rebirth, memories, habits or tendencies are carried over from one life to another. This transfer of memory is part of the action of *karma*, a kind of causation across time; actions bring about effects in the future, even in later lives. In Christianity there are several different theories of survival, but all imply a survival of memory. According to the Roman Catholic doctrine of Purgatory, after death believers enter an ongoing process of development, comparable to dreaming. This process would make no sense unless the person's memories played a part in the process. Some Protestants believe that after death everyone goes to sleep, only to be resurrected just before the Last Judgment. But this theory too requires a survival of memory because the Last Judgment would be meaningless if the person being judged had forgotten who he was and what he had done.

By contrast, the materialist theory is simple. Memories are in the brain; the brain decays at death; therefore all memories are wiped out for ever. For an atheist, what could be a better proof of the folly of religious belief? All religious theories of survival are impossible because they all rely on the survival of personal memories, which are wiped out when the brain decays. The materialist theory leaves the question of the survival of bodily death closed. By contrast, the resonance theory leaves the question open. Memories themselves do not decay at death, but can continue to act by resonance, as long as there is a vibratory system that they can resonate with. They contribute to the collective memory of the species. But whether or not there is an immaterial part of the self that can still access these memories in the absence of a brain is another question.

Questions for materialists

Do you believe that memories are stored as material traces in brains? If so, can you summarise the evidence?

How do you think memory-retrieval systems recognise the memories they are trying to retrieve from memory stores?

Have you ever considered the possibility that memory might depend on some kind of resonance rather than on material traces?

If the trace theory of memory is a testable hypothesis, rather than a dogma, how could it be established experimentally that memory depends on traces rather than resonance?

SUMMARY

Repeated failures to find memory traces fit well with the idea of memory as a resonant phenomenon, where similar patterns of activity in the past affect present activities in minds and brains. Individual and collective memory both depend on resonance, but self-resonance from an individual's own past is more specific and hence more effective. Animal and human learning may be transmitted by morphic resonance across space and time. The resonance theory helps account for the ability of memories to survive serious damage to brains, and is consistent with all known kinds of remembering. This theory predicts that if animals, say, rats, learn a new trick in one place, say, Harvard, rats all over the world should be able to learn it faster thereafter. There is already evidence that this actually happens. Similar principles apply to human learning. For example, if millions of people take standard tests, like IQ tests, the tests should become progressively easier, on average, for other people to do. Again, this seems to be what happens. Individual memory and collective memory are different aspects of the same phenomenon and differ in degree, not in kind.

8

Are Minds Confined to Brains?

Materialism is the doctrine that only matter is real. Hence minds are in brains, and mental activity is nothing but brain activity. This assumption conflicts with our own experience. When we look at a blackbird, we see a blackbird; we do not experience complex electrical changes in our brains. But most of us accepted the mind-within-the-brain theory before we ever had a chance to question it. We took it for granted as children because it seemed to be supported by all the authority of science and the educational system.

In his study of children's intellectual development, the Swiss psychologist Jean Piaget found that before about the age of ten or eleven, European children were like 'primitive' people. They did not know that the mind was confined to the head; they thought it extended into the world around them. But by about the age of eleven, most had assimilated what Piaget called the 'correct' view: 'Images and thoughts are situated in the head.'[1]

Educated people rarely question this 'scientifically correct' view in public, perhaps because they do not want to be thought stupid, childish or primitive. Yet the 'correct' view conflicts with our most immediate experience every time we look around us. We see things outside our bodies; we do not experience images inside our heads. The materialist theory dominated academic psychology for most of the twentieth century. The long-dominant behaviourist school explicitly denied the reality of consciousness. The leading American behaviourist, B. F. Skinner, proclaimed in 1953 that mind and consciousness were non-existent entities 'invented for the sole purpose of providing spurious explanations . . . Since mental or psychic events are asserted to lack the dimensions of physical science, we have an additional reason for rejecting them.'[2] As discussed in Chapter 4, a similar denial of conscious experience is still advocated by contemporary philosophers of the school known as 'eliminative materialism'. Paul Churchland, for example, argues that subjectively experienced mental states should be regarded as non-existent because descriptions of such states cannot be reduced to the language of neuroscience.[3]

Likewise, many leading scientists regard conscious experience as nothing but the subjective experience of brain activity (see Chapter 4). Francis Crick called this the Astonishing Hypothesis:

> 'You', your joys and your sorrows, your memories and your ambitions, your sense of personal identity and free will, are in fact no more than the behaviour of a vast assembly of nerve cells and their associated molecules . . . This hypothesis is so alien to the ideas of most people alive today that it can truly be called astonishing.[4]

This is, indeed, an astonishing claim. But within institutional science it is commonplace. Crick was no revolutionary: he spoke for the mainstream. Susan Greenfield, an influential neuroscientist, looked at an exposed brain in an operating theatre and reflected, 'This was all there was to Sarah, or indeed to any of us . . . We are but sludgy brains, and . . . somehow a character and a mind are generated in this soupy mess.'[5]

The traditional alternative to materialism is dualism, the doctrine that minds and brains are radically different: minds are immaterial and brains are material; minds are outside time and space, matter is inside time and space. Dualism makes better sense of our experience but makes no sense in terms of mechanistic science, which is why materialists reject it so vehemently (see Chapter 4).

We need not stay stuck in this materialist-dualist contradiction. There is a way out: a field theory of minds. We are used to the fact that fields exist both within and outside material objects. The field of a magnet is inside it and also extends beyond its surface. The gravitational field of the earth is inside the earth and also stretches out far beyond it, keeping the moon in its orbit. The electromagnetic field of a mobile phone is both inside it and extends all around it. In this chapter I suggest that the fields of minds are within brains and extend beyond them.

Extended minds

If we follow Francis Crick and treat materialism as a hypothesis, rather than a philosophical dogma, it should be testable. As Carl Sagan liked to say, 'Extraordinary claims demand extraordinary evidence.' Where is the extraordinary evidence for the materialist claim that the mind is nothing but the activity of the brain?

There is very little. No one has ever seen a thought or image inside someone else's brain, or inside his or her own.[6] When we look around us, the images of the things we see are outside us, not in our heads. Our experiences of our bodies are in our bodies. The feelings in my fingers are in my fingers, not in my head. Direct experience offers no support for the extraordinary claim that all experiences are inside brains. Direct experience is not irrelevant to the nature of consciousness: it *is* consciousness.

Extended minds are implicit in our language. The words 'attention' and 'intention' come from the Latin root *tendere*, to stretch, as in 'tense' and 'tension'. 'Attention' is *ad* + *tendere*, 'to stretch towards'; 'intention', *in* + *tendere*, 'to stretch into'.

How does vision work?

A debate about the nature of vision was going on in ancient Greece 2,500 years ago. It was taken up in the Roman Empire and in the Islamic world, and continued in Europe throughout the Middle Ages and the Renaissance. The debate played an important part in the birth of modern science, and is still alive today.

There were three main theories of how we see. The first was that vision involves an outward projection of invisible rays through the eyes. This is often called the 'extramission' theory, which literally means 'sending out'. Second was the idea of a 'sending in' of images through light into the eyes, the 'intromission' theory. The third theory, a combination of the other two, states that there is both an inward movement of light and an outward movement of attention.

The extramission theory agrees with people's experience of vision as an active process. We look *at* things, and can decide where to direct our attention. Vision is not passive. Plato supported this theory of vision, and around 300 BC Euclid, famous for his works on geometry, worked it out in mathematical detail. He showed how projection of virtual images from the eye could explain how we see images in mirrors. Unlike light itself, which is reflected by mirrors, visual projections go straight through them. They are not material.

Isaac Newton accepted Euclid's theory, and illustrated it in 1704 in his book *Opticks* (Figure 8.1). Essentially the same diagram is used in science textbooks today. A typical British physics textbook for secondary schools describes the process as follows: 'Rays from a point on the object are reflected at the mirror and appear to come from a point behind the mirror where the eye imagines the rays intersect when produced backwards.'[7] There is no discussion of how the eye 'imagines' rays intersecting, or how it produces them backwards. This is essentially Euclid's extramission theory of virtual images, but its implications are left implicit.

Since the early seventeenth century the intromission theory has been scientifically orthodox, largely thanks to the work of Johannes Kepler (1571–1630), best known for his discoveries in astronomy. Kepler realised that light entering the eye through the pupil was focused by the lens, and produced an inverted image on the retina. He published his theory of the retinal image in 1604. Although this was a major triumph, and a landmark in the development of modern science, it raised questions that Kepler could not answer, and are still unanswered today. The problem was that the images on the retinas of both eyes were inverted and reversed; in other words, they were upside down

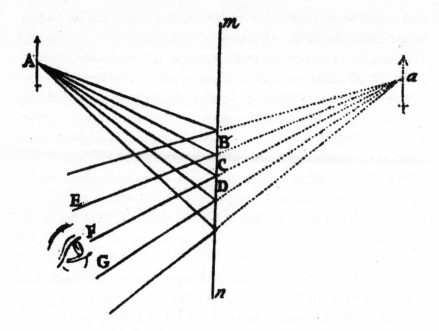

FIGURE 8.1. Isaac Newton's diagram of reflection in a plane mirror: 'If an Object A can be seen by Reflexion of a Looking-glass mn, it shall appear, not in its proper place A, but behind the Glass at a.' (Newton, 1704, Fig. 9)

and the left side was at the right, and vice versa. Yet we do not see two small, inverted, reversed images.[8]

The only way Kepler could deal with this problem was by excluding it from optics. Once the image had been formed on the retina, it was someone else's business to explain how we actually see it.[9] Vision itself was 'mysterious'. Ironically, the triumph of the intromission theory was achieved by leaving the experience of seeing unexplained. This problem has haunted science ever since.

Kepler's contemporary, Galileo Galilei (1564–1642), likewise withdrew perceptions from the external world and squeezed them into the brain. He made a distinction between what he called primary and secondary qualities of objects. The primary qualities were those that could be measured and treated mathematically, such as size, weight and shape. These were the concern of objective science. The secondary qualities, such as colour, taste, texture and smell, were not within matter itself. They were subjective rather than objective. And subjective meant within the brain. Thus our direct experience of the world was split into two separate poles, the objective, out there, and the subjective, within the brain.

After four hundred years of mechanistic science, there has been almost no progress in understanding how the brain produces subjective experience, although many details have been discovered about the activities of different regions of the brain. The orthodox assumption is that the brain constructs a picture or model of the world inside itself. This is how an authoritative textbook called *Essentials of Neural Science and Behavior* described the process:

> [T]he brain constructs an internal representation of external physical events after first analyzing them into component parts. In scanning the visual field the brain simultaneously but separately analyzes the form of objects, their movement, and their color, all before putting together an image according to the brain's own rules.[10]

Most contemporary metaphors for the activity of the brain are derived from computers, and 'internal representations' are commonly conceived of as 'virtual reality' displays. As the psychologist Jeffrey Gray put it succinctly, 'The "out there" of conscious experience isn't really out there at all; it's inside the head.' Our visual perceptions are a 'simulation' of the real world that is 'made by, and exists within, the brain'.[11]

The idea of visual experiences as simulations inside heads leads to strange consequences, as the philosopher Stephen Lehar has pointed out.[12] It means that when I look at the sky, the sky I see is inside my head. My skull is beyond the sky!

> I propose that out beyond the farthest things you can perceive in all directions, i.e. above the dome of the sky, and below the solid earth under your feet, or beyond the walls and ceiling of the room you see around you, is located the inner surface of your

true physical skull, beyond which is an unimaginably immense external world of which the world you see around you is merely a miniature internal replica. In other words, the head you have come to know as your own is not your true physical head, but only a miniature perceptual copy of your head in a perceptual copy of the world, all of which is contained within your real head.[13]

Despite the theories of academic scientists and philosophers, most people do not accept that all their experiences are located inside their heads. They think they are where they seem to be, outside their heads.

In the 1990s, Gerald Winer and his colleagues in the psychology department at Ohio State University investigated people's beliefs about the nature of vision through a series of questionnaires and tests. They were surprised that extramission beliefs were common among children, and 'shocked' when they discovered that they were also widespread among college students, even among those studying psychology, who had been taught the 'correct' theory of vision.[14] Among schoolchildren from grades five to eight, more than 70 per cent believed in a combined intromission-extramission theory, and among college students 59 per cent.[15] Winer and his colleagues called this a 'striking instance of a scientific misconception'.[16] Education had failed to convert most of the students to the correct belief:

> Given that extramissionists in our studies affirm extramission even though they have been taught about vision, our attention is now directed to understanding whether education can eradicate these odd, but seemingly powerful, intuitions about perception.[17]

Winer and his colleagues seem doomed to failure in their crusade for intellectual cleansing. These 'odd' intuitions about perception persist because they are closer to experience than the official doctrine, which leaves so much unexplained – including consciousness itself.

Images outside bodies

Not all philosophers and psychologists believe the mind-in-the-brain theory, and over the years a minority has always recognised that our perceptions may be just where they seem to be, in the external world outside our heads, rather than representations inside our brains.[18] In 1904, William James wrote:

> [T]he whole philosophy of perception from Democritus' time downwards has been just one long wrangle over the paradox that what is evidently one reality should be in two places at once, both in outer space and in a person's mind. 'Representative' theories of perception avoid the logical paradox, but on the other hand they violate the reader's sense of life which knows no intervening mental image but seems to see the room and the book immediately as they physically exist.[19]

As Alfred North Whitehead expressed it in 1925, 'sensations are projected by the mind so as to clothe appropriate bodies in external nature'.[20]

A recent proponent of the extended mind is the psychologist Max Velmans. In his book *Understanding Consciousness* (2000), he proposed a 'reflexive model' of the mind, which he illustrated by this discussion of a subject (S) looking at a cat:

> According to reductionists there seems to be a phenomenal cat 'in S's mind', but this is really nothing more than a state of her brain. According to the reflexive model, while S is gazing at the cat, her only visual experience of the cat is the cat she sees out in the world. If she is asked to point to this phenomenal cat (her 'cat experience'), she should point not to her brain but to the cat as perceived, out in space beyond the body surface.[21]

Velmans suggested that this image might be like 'a kind of neural "projection hologram"'. A projection hologram has the interesting quality that the three-dimensional image it encodes is perceived to be out in space, in *front* of its two-dimensional surface.'[22] But Velmans was ambiguous about the nature of this projection. A hologram is, after all, a field phenomenon. He called it 'psychological' rather than 'physical' and in the end said he did not know how it happened, but added, 'not fully understanding *how* it happens does not alter the fact *that* it happens'.

My own suggestion is that the outward projection of visual images is both psychological *and* physical. It occurs through perceptual fields. These are psychological, in the sense that they underlie our conscious perceptions, and also physical or natural in that they exist outside the brain and have detectable effects. Human perception is not unique in being extended through seeing and hearing. Other animals see things through fields projected beyond the surfaces of their bodies, and hear things through projected auditory fields. We are like other animals.

The senses are not static. The eyes move as we look at things, and our heads and entire bodies move around in our environments. As we move, our perceptual fields change. Perceptual fields are not separate from our bodies, but include them. We can see our own outer surface, our skin, hair and clothing. We are inside our fields of vision and action. Our awareness of three-dimensional space includes our own bodies within it, and our movements and intentions in relation to what is around us. Like other animals, we are not passive perceivers but active behavers, and our perceptions and behaviour are closely linked.[23]

Some neuroscientists and philosophers agree that perceptions depend on the close connection between perception and activity, linking an animal or person to the environment. One school of thought advocates an 'enactive' or 'embodied' or 'sensorimotor' approach. Perceptions are not represented in a world-model inside the head, but are enacted or 'brought forth' as a result of the interaction of the organism and its environment. As Francisco Varela and his colleagues expressed it, 'perception and action have evolved together . . . perception is always *perceptually guided activity*'.[24] As the philosopher Arva Noë put it, 'We are out of our heads. We are in the world and of it. We are patterns of active engagement with fluid boundaries and changing components. We are distributed.'[25] The psychologist Kevin O'Regan, a committed materialist, prefers this approach to the mind-in-the-brain theory precisely because he wants to expel all magic from the brain. He does not accept that seeing is in the brain, because this would 'put you in the terrible situation of having to postulate some magical mechanism that endows the visual cortex with sight, and the auditory cortex with hearing'.[26]

Henri Bergson anticipated the enactive and sensorimotor approaches more than a century ago. He emphasised that perception is directed towards action. Through perception, 'The objects which surround my body reflect its possible action upon them.'[27] The images are not inside the brain:

> The truth is that the point P, the rays which it emits, the retina and the nervous elements affected, form a single whole; that the luminous point P is a part of this whole; and that it is really in P, and not elsewhere, that the image of P is formed and perceived.[28]

My own interpretation is that vision takes place through extended perceptual fields, which are both within the brain and stretch out beyond it.[29] Vision is rooted in the activity of the brain, but is not confined to the inside of the head. Like Velmans, I suggest that the formation of these fields depends on changes in various regions of the brain as vision takes place, influenced by expectations, intentions and memories. These are a kind of morphic field and, like other morphic fields, connect together parts within wholes, and have an inherent memory given by morphic resonance from similar fields in the past (see Chapter 3). When I look at a person or an animal, my perceptual field interacts with the field of the person or animal I am looking at, enabling my gaze to be detected.

Our experience certainly suggests that our minds are extended beyond our brains. We see and hear things in the space around us. But there is a strong taboo against anything that suggests that seeing and hearing might involve any kind of outward projection. This issue cannot be resolved by theoretical arguments alone, or else there would have been more progress over the last century – or even over the last 2,500 years.

I am convinced that the way forward is to treat fields of the mind as a testable scientific hypothesis rather than a philosophical theory. When I look at something, my perceptual fields 'clothe' what I am looking at. My mind touches what I am seeing. Therefore I might be able to affect another person just by looking. If I look at someone from behind when she cannot hear me, or see me, and does not know I am there, can she feel my gaze?

The detection of stares

Most people have felt someone looking at them from behind, turned around and met the person's eyes. Most people have also experienced the converse: they have sometimes made people turn round by staring at them. In extensive surveys in Europe and North America, between 70 and 97 per cent of adults and children reported experiences of these kinds.[30]

In surveys I carried out in Britain, Sweden and the United States, these experiences seemed to be most common when people were being stared at by strangers in public places, such as streets and bars. They happened more when people felt vulnerable than when they felt secure.

When people made others turn around by staring at them, both men and women said that curiosity was their most frequent reason for staring, followed by a desire to attract the other person's attention. Other motives included sexual attraction, anger and affection.[31] In short, the ability to detect someone's attention was associated with a range of motives and emotions.

In some Oriental martial arts, students are trained to increase their sensitivity to being looked at from behind.[32] And some people observe others for a living. The sense of being stared at is well known to many police officers, surveillance personnel and soldiers, as I found through a series of interviews with professionals. Most felt that some people they were watching seemed to know, even though the watchers were well hidden. For example, a narcotics officer in Plains, Texas, said, 'I've noticed that a lot of times the crook will just get a feeling that things aren't right, that he's being watched. We often have somebody look right in our direction even though he can't see us. A lot of times we're inside a vehicle.' When detectives are trained to follow people, they are told not to stare at their backs any more than necessary because otherwise the person might turn around, catch their eye and blow their cover.[33]

According to experienced surveillance officers, this sense also works at a distance when people are watched through binoculars. Several soldiers told me that some people could tell when they were being looked at through telescopic sights. For example, a soldier in the US Marine Corps served as a sniper in Bosnia in 1995, where he was assigned to shoot 'known terrorists'. While aiming through the telescopic sight of his rifle, he found that people seemed to know when he was aiming at them. 'Within one second prior to actual termination, a target would somehow seem to make eye contact with me. I am convinced that these people somehow sensed my presence at distances over one mile. They did so with uncanny accuracy, in effect to stare down my own scope.'

Many celebrity photographers have had similar experiences. One long-lens photographer who worked for the *Sun*, the most popular tabloid newspaper in Britain, said that he was amazed by how many times his quarries would 'turn round and look right down the lens', even if they were looking in the opposite direction to start with. He did not think they could see him or detect his movements. 'I am talking about taking pictures at distances of up to half a mile away in situations where it is quite impossible for people to see me, although I can see them. They are so aware it is uncanny.'[34]

Many species of non-human animals also seem able to detect looks. Some hunters and wildlife photographers are convinced that animals can detect their gaze even when they are hidden and looking at the animals through telescopic lenses or sights. One British deer hunter found that the animals seemed to detect his intention, especially if he delayed shooting when he had them in his rifle sights: 'If you wait a fraction too long, it will just take off. It'll sense you.'

Several bird photographers told me that when they were in hides, invisible to the birds they were watching, the birds still seemed to know when they were being looked at. One said, 'I spend a lot of time in hides and it is uncanny how birds can just seem to sense you are there, become agitated, even though you know you haven't moved. With herons you can tell instantly that they are alert to danger. Very often the lens is completely still and they suddenly seem to realise that there is something looking at them, and their heads go up and they go very stiff and wait to see if they can see anything else.'[35]

Conversely, some photographers and hunters had felt wild animals looking at them.[36] The naturalist William Long wrote that when he was sitting in the woods alone,

> I often found within myself an impression which I expressed in the words, 'Something is watching you.' Again and again, when nothing stirred in my sight, that curious warning would come; and almost invariably, on looking around, I would find some bird or fox or squirrel which had probably caught a slight motion of my head and had halted his roaming to creep near and watch me inquisitively.[37]

Some pet owners claim that they can wake their sleeping dogs or cats by staring at them. Others have found it works the other way round and that their animals can wake them by staring.

In their surveys in Ohio, Winer and his colleagues found that more than a third of their respondents said they had felt when animals were looking at them. About half believed that animals could feel their looks, even when the animals could not see their eyes.[38]

If the sense of being stared at is real, then it must have been subject to evolution by natural selection. How might it have evolved? The most obvious possibility is in the context of predator-prey relations. Prey animals that detected when predators were looking at them would stand a better chance of surviving than those that did not.[39]

Experimental tests

Since the 1980s the sense of being stared at has been investigated experimentally both through direct looking and also through closed circuit television (CCTV). In the scientific literature it is variously referred to as 'unseen gaze detection' or 'remote attention' or 'scopaesthesia' (from Greek *skopein*, to view, and *aisthetikos*, sensitive).

In direct-looking experiments, people work in pairs, with a subject and a looker. In a randomised series of trials, blindfolded subjects sit with their backs to the lookers, who either stare at the back of their necks, or look away and think of something else. A mechanical signal – a click or a beep – marks the beginning of each trial. Within a few seconds the subjects guess whether they are being looked at or not. Their guesses are either right or wrong, and are recorded immediately. A test usually consists of twenty trials.

These tests are so simple that a child can do them, and thousands of children already have. In the 1990s, this research was popularised through *New Scientist* magazine, BBC TV and Discovery Channel TV, and many tests were conducted in schools and as student projects at universities. Altogether, tens of thousands of trials were carried out.[40] The results were remarkably consistent. Typically, about 55 per cent of the guesses were right, as opposed to 50 per cent expected by chance. Although the effect was small, because it was so widely replicated it was highly significant statistically. In more rigorous experiments subjects and starers were separated by windows or one-way mirrors, eliminating the possibility of subtle cues by sound or even smell. They were still able to tell when they were being watched.[41]

The largest experiment on the sense of being stared at began in 1995 at the NEMO Science Centre in Amsterdam. More than 18,000 pairs took part, with positive results that were highly significant statistically.[42] The most sensitive subjects were children under the age of nine.[43]

Surprisingly, the sense of being stared at works even when people are looked at on screens, rather than directly. CCTV systems are routinely used for surveillance in shopping malls, banks, airports, streets and other public spaces. My assistants and I interviewed surveillance officers and security personnel whose job it was to observe people on screens. Most were convinced that some people could feel when they were being watched.[44] The security manager in a large firm in London had no doubt that some people have a sixth sense: 'They can have their backs to the cameras, or be scanned using hidden devices, yet they still become agitated when the camera is trained on them. Some move on, some look around for the camera.'

In laboratory tests, many people respond physiologically to being watched through CCTV, even though they are unconscious of their response. In these experiments, the researchers put a subject in one room and a looker in another, where the subject could be watched through CCTV. The subjects' galvanic skin response was recorded, as in lie-detector tests, enabling emotional changes to be detected through differences in sweating; wet skin conducts electricity better than dry skin. In a randomised series of trials, the starers either looked at the subject's image on the TV monitor, or looked away and thought of something else. The subjects' skin resistance changed significantly when they were being looked at.[45]

The fact that gaze detection works through CCTV shows that people can detect other people's attention even when they are not being watched directly.

The effects of attention at a distance show that minds are not confined to the insides of brains.

Minds extended in time

Minds extend beyond brains in time as well as space. We are connected to the past by memory and habit, and to the future by desires, plans and intentions. Are these memories and virtual futures contained materially within brains in the present, or are minds connected to the past and future by non-material links?

The conventional answer is that our memories and intentions must be inside brains in the present. Where else could they be? The computer metaphor reinforces this way of thinking. A computer's memories are stored on magnetic or optical disks, or in solid-state memory systems. These memories are material structures or patterns in the present. And just as the computer's memories exist physically in its present, so its programmed goals are present in it too. Past and future are both physically present. By analogy, memories, goals, plans and intentions are physically present in brains.

The assumption that memories are stored materially inside brains was discussed in the previous chapter. The assumption that future goals are inside brains is equally questionable. They exist in a realm of possibility; they are virtual futures. Possibilities are not material. In quantum physics, the wave function that describes how electrons or other particles might behave is a mathematical model in a multi-dimensional space based on 'complex numbers' that include an imaginary number, the square root of -1. The wave function maps possible future states of the system in terms of probabilities. When a quantum particle such as an electron interacts with a physical system, for example in a process of being measured in a laboratory, the wave function collapses into one of its many possible outcomes. Many possibilities are reduced to an objectively observable fact, just as they are when a person takes a decision and acts on it. But the wave function itself is not material; it is a mathematical description of possibilities.

As the philosopher Alfred North Whitehead suggested, minds and matter are related as processes in time, rather than in space (see Chapter 4). The subject chooses among its potential futures, and the direction of mental causation runs from potential futures to the present. Neither the future nor the past is material, but both have effects in the present through memories, habits and choices.

According to the hypothesis of morphic resonance, similar processes occur at all levels of organisation, including biological morphogenesis. As a carrot seed develops into a carrot plant, it is shaped by its morphogenetic fields, inherited from previous carrot plants by morphic resonance. These morphogenetic fields contain the attractors and chreodes that channel its development towards the form of a mature plant (see Chapters 5 and 6). Neither inherited habits nor future goals are material structures present in the plant; instead they are patterns of goal-directed activity. In a similar way neither memories nor purposes are contained in brains, although they influence brain activity.

Most of our mental activity is habitual and unconscious. Conscious mental activity is largely concerned with possible actions, including speaking. Our conscious minds inhabit the realm of possibility, and languages greatly expand the possibilities they can entertain. Think of hearing a story. Our minds can embrace possibilities that go far beyond our own experience. Conscious minds choose among possibilities, and their choices collapse possibilities into actions that are objectively observable in the physical world. The arrow of causation is from the virtual future, going 'backwards' in time. In this sense minds act as final causes, setting goals and purposes.

In order to make choices, minds must contain alternative possibilities: coexisting at the same time. In the language of quantum physics, these possibilities are 'superposed'. The physicist Freeman Dyson wrote, 'The processes of human consciousness differ only in degree but not in kind from the processes of choice between quantum states which we call "chance" when made by electrons.'[46]

According to the hypothesis of morphic resonance, all self-organising systems, including protein molecules, *Acetabularia* cells, carrot plants, human embryos and flocks of birds, are shaped by memory from previous similar systems transmitted by morphic resonance and drawn towards attractors through chreodes. Their very being involves an invisible presence of both past and future. Minds are extended in time not because they are miraculously different from ordinary matter, but because they are self-organising systems. All self-organising systems are extended in time, shaped by morphic resonance from the past, and drawn towards attractors in the future.

What difference does it make?

Liberating minds from confinement in heads is like being released from prison. Most people have already broken out in secret. Even most materialists are not true believers when it comes to themselves; they effectively ignore the materialist theory in their private lives. They do not take seriously the idea that their skulls are beyond the sky. In practice, they are dualists who believe they make free choices.

Those who take their materialist faith seriously ought to believe that they are like robots with no free will. And some materialists actually *want* to experience themselves as automata. For example, the psychologist Kevin O'Regan told his fellow materialist Susan Blackmore, 'Ever since I've been a child I've wanted to be a robot. I think one of the great difficulties of human life is that one's life is inhabited by uncontrollable desires and that if one could only be master of those and become more like a robot one would be much better off.' He thought everyone else was a robot too, but 'just labouring under the illusion that they weren't'. But as Blackmore pointed out, a robot with emotions it could control would be an unusual kind of robot.[47] O'Regan is exceptional in extending materialist theories to the realm of private life, but nevertheless he endowed his robot-self with a desire to be master of his emotions, implying both conscious experience and choice.

Materialism is unpersuasive if one takes one's own experience into account. But because it is the creed of established science, its authority is enormous. That is why so many educated people try to resolve this dilemma by adopting a materialist persona in scientific discourse, while in private accepting the reality of conscious experience and choice.

A field theory of minds and bodies liberates us from this stalemate. Minds are closely connected to fields that extend beyond brains in space, and also extend beyond brains in time, linked to the past by morphic resonance and to virtual futures through attractors.

Questions for materialists

When you look at the sky, do you think that the sky you are seeing is inside your skull, and that your skull is beyond the sky?

Have you ever felt someone was looking at you from behind, or have you ever made someone turn around by staring at him?

Do you believe that all your conscious life and all your bodily experience is inside your brain?

In quantum physics, electrons are described by wave equations that include all the electron's future possibilities, which are not material. Do you think that the possibilities among which you choose are more material than those of electrons?

SUMMARY

Our minds are extended in every act of perception, reaching even as far as the stars. Vision involves a two-way process: the inward movement of light into the eyes, and the outward projection of images. What we see around us is in our minds but not in our brains. When we look at something, in a sense our mind touches it. This may help to explain the sense of being stared at. Most people say they have felt someone looking at them from behind, and claim to have made people turn round by looking at them. The ability to detect stares seems to be real, as shown in many scientific tests, and even seems to work through closed circuit television. Minds are extended beyond brains not only in space but also in time, and connect us to our own pasts through memory and to virtual futures, among which we choose.

9

Are Psychic Phenomena Illusory?

Most materialist dogmas go almost unquestioned. But the claim that psychic phenomena are illusory is inevitably controversial. Most people have had seemingly telepathic or precognitive experiences. Most have felt they were being looked at from behind, and have turned round, or they have made others turn by staring at them, as discussed in Chapter 8. Many pet owners have noticed that their dogs and cats seem to pick up their intentions, even when they are out of sight. Such occurrences are sometimes called intuitive, or psychic, or parapsychological, or are attributed to a sixth or seventh sense, or to extra-sensory perception (ESP) or to psi, short for psyche.

For committed materialists, all such phenomena are illusory. Minds are inside brains, and mental activity is nothing but electrochemical brain activity. Hence thoughts and intentions cannot have direct effects at a distance; neither can minds be open to influences from the future. Although such paranormal phenomena *seem* to occur, they must have normal explanations in terms of coincidence, or subtle sensory cues, or wishful thinking, or fraud.

This controversy has endured for generations and raises the question of the nature of science itself. Is science a belief system, or is it a method of enquiry? Materialism has been the standard view since the late nineteenth century, but a small minority of researchers has continued to investigate psychic phenomena because they will enlarge our understanding of minds, if they are real, and widen the scope of science.

The first organisation devoted to the investigation of these phenomena, the British Society for Psychical Research, was established in 1882. Its purpose is still reprinted in every issue of the *Journal of the Society for Psychical Research*: 'To examine without prejudice or presupposition and in a scientific spirit those faculties of man, real or supposed, which appear to be inexplicable on any generally recognised hypothesis.' From the very beginning this enterprise was controversial. Commenting on this new organisation, the physiologist Hermann von Helmholtz, who played such an important part in the establishment of the principle of conservation of energy in living organisms (see Chapter 2), dismissed the possibility of telepathy out of hand: 'Neither the testimony of all the fellows of the Royal Society, nor even the evidence of my own sense, would lead me to believe in the transmission of thought from one person to another independently of the recognised channels of sense. It is clearly impossible.'[1]

Not much has changed. Although a growing body of evidence from psychic research and parapsychology suggests that telepathy, precognition and other psychic phenomena are real, materialists still believe they are impossible and that psychical research is inherently pseudoscientific. In 2010, the veteran sceptic James Alcock declared:

> The parapsychological quest is motivated not by scientific theory, nor by anomalous data produced in the course of mainstream science. Rather, it is motivated by deeply held beliefs on the part of the researchers – belief that the mind is more than an epiphenomenal reflection of the physical brain, belief that it is capable of transcending the physical limits normally imposed by time and space. It is this belief in the possibility of such impossible things that sustains parapsychology and leaves it relatively undaunted by the slings and arrows of (yes, sometimes outrageous) criticism.[2]

This situation recalls a story about a witty English clergyman, Sydney Smith. Around 1800, he was strolling along a narrow street with a friend when they passed beneath two women leaning out of opposite windows screaming insults at each other. 'The two ladies will never agree,' Smith commented, as the debate raged over his head, 'for they are arguing from different premises.'[3]

The materialist premise is that the nature of minds is already understood in principle: mental activity is brain activity, and is located inside heads. Hence psychic phenomena are impossible. The premise of psychic researchers is that psychic phenomena are possible, although not yet understood; and only by studying them can more be found out.

These different premises are also reflected in the terms 'normal' and 'paranormal'. Psychic phenomena are normal in the sense that they are common: for example, most people have made other people turn round by staring at them, or had seemingly telepathic experience with telephone calls, as discussed below. But because these experiences do not fit in with the materialist mind-in-brain theory, they are classified as paranormal, literally meaning 'beyond the normal'. In this sense, 'normal' is defined not by what actually happens, but by the assumptions of materialists.

Likewise, the term parapsychology means 'beyond psychology' and implies that it is not part of normal psychology. I think this term is unfortunate and prefer the older term 'psychic research', or 'psi research' for short. If psi phenomena exist, which I think they do, they are normal, not paranormal; natural, not supernatural. They are part of human nature and animal nature, and they can be investigated scientifically.

Sceptics often repeat the slogan that 'extraordinary claims demand extraordinary evidence', which is another expression of the materialist assumption. The sense of being stared at and telepathy are ordinary, in that most people have experienced them. They are not 'extraordinary', meaning 'beyond the normal order', or 'highly exceptional':[4] they are common. From this point of view, the *sceptics'* claim is extraordinary and demands extraordinary evidence. Where is the extraordinary evidence that most people are deluded about their own experience? Sceptics can only fall back on generic arguments about the fallibility of human judgement – or, rather, other people's judgements.

In this chapter, I consider research on telepathy and precognition or presentiment. In the interests of brevity, I omit the other two main areas of psi research: clairvoyance, the ability to see or experience things at a distance, sometimes called remote viewing; and psychokinesis, or mind-over-matter effects.[5] I then return to the opinions of sceptics.

How an open-minded scientist opened my mind

Telepathy literally means 'distant feeling', from the Greek *tele*, 'distant', as in telephone and television, and *pathe*, 'feeling', as in sympathy and empathy.

In the course of my scientific education at school and university, I was converted to the materialist worldview, and absorbed the standard attitude towards telepathy and other psychic phenomena. I dismissed them. I did not study the evidence because I assumed there was none worth reading. But when I was a graduate student in the Department of Biochemistry at Cambridge University, someone mentioned telepathy in a conversation in the laboratory tearoom. I dismissed it out of hand. But sitting nearby was one of the doyens of British biochemistry, Sir Rudolph Peters, formerly professor of biochemistry at Oxford, who after retirement continued his research in our laboratory in Cambridge. He was kindly, his eyes twinkled, and he had more curiosity than most people half his age. He asked if any of us had ever looked at the evidence. We had not. He told us he had done some research on this subject himself, and had come to the conclusion that something unexplained was really happening. He later told me the story in detail, and gave me a paper he had published on the subject in the *Journal of the Society for Psychical Research*.[6]

A friend of his, E. G. Recordon, an ophthalmologist, had a boy patient who was severely disabled, mentally retarded and almost blind. Yet in routine eye tests he seemed able to read the letters very well, apparently by 'remarkable guesswork'. Recordon said, 'It gradually dawned on me that this "guesswork" was particularly interesting; and I came to the conclusion that he must be working through his mother.' It turned out that the boy could only read the letters when his mother was looking at them, raising the possibility of telepathy.

Peters and Recordon did some preliminary experiments at the family's home. A screen separated the mother and son, preventing the boy from picking up any visual cues. When his mother was shown a series of written numbers or words, the boy guessed many of them correctly. Peters and his colleagues could not observe any sign of cueing by sound or subtle movements. They then carried out two experiments over the telephone, which they tape-recorded. The mother was taken to a laboratory six miles away, while the boy remained at home in Cambridge. The experimenters had a set of cards on which randomly selected numbers or letters were written. The cards were shuffled so that they were in a random order. One of the researchers turned up a card and showed it to the mother. The boy, at the other end of the telephone line, then guessed what it was, and the mother responded by saying 'right' or 'no'. The mother was then shown the next card, and so on. Each trial lasted only a few seconds.

In the trials with letters, there was a 1 in 26 (3.8 per cent) chance of guessing the letter correctly at random. The boy guessed correctly in 38 per cent of the trials. When he was wrong, he was given a second guess and was right 27 per cent of the time. In experiments with random numbers he was likewise correct far more than would have been expected by random guessing. The odds against these results arising by chance were billions to one. Peters concluded that this was indeed a case of telepathy, which had developed to an unusual degree because of the boy's extreme needs and the mother's desire to help him.[7] As he remarked, 'In every respect the mother was emotionally involved in trying to help her backward son.'

As I later came to understand, telepathy usually occurs between people who are closely bonded, such as parents and children, spouses, and close friends.[8] Peters's investigation was unusual in that he studied a case where the bonds between the 'sender' and the 'receiver' were unusually strong. By contrast, most experiments by psychic researchers and parapsychologists have used pairs of strangers, between whom the effects were much smaller. Nevertheless, taken together, these experiments produced an impressive body of evidence.

Telepathy in the laboratory

Between 1880 and 1939, dozens of investigators published a total of 186 papers describing four million card-guessing trials, in which subjects guessed which randomly selected card a 'sender' was looking at. Most of these tests gave hit rates modestly above the level expected by random guessing, and when combined together in a statistical procedure called meta-analysis, the overall results were hugely significant statistically.[9]

Sceptics often argue that such impressive collections of data are misleading, because researchers may publish only positive results, and leave negative studies unpublished in their files, the so-called 'file-drawer effect'. This objection is plausible, but it applies to all branches of science, including physics, chemistry and biology, where most data go unpublished. Psychic researchers experience much more sceptical scrutiny than scientists in conventional fields, and are also more aware of the importance of publishing statistically non-significant results, and actually do so. Anyway, calculations show how many unsuccessful studies would be needed to bring down the results of these card-guessing tests to chance levels. There would have to be 626,000 unpublished reports, or in other words 3,300 unpublished studies for every one that was published.[10] This is implausible.

Many card-guessing trials were carried out at the famous parapsychology laboratory at Duke University, North Carolina, from the 1920s to the 1960s using a set of five specially prepared cards with abstract symbols on them. By chance subjects would have been right 20 per cent of the time. In hundreds of thousands of trials, the average hit rate was 21 per cent, only slightly above the chance level, but highly significant statistically because of the large number of trials.[11]

Unfortunately, the experimenters' desire to follow rigorous scientific procedures led them to adopt procedures far removed from the way telepathy occurs in real life. These repetitive, boring laboratory tests between strangers using abstract stimuli were about as unnatural as it was possible to be.

In the 1960s, a new generation of researchers tried to find ways of doing research that were closer to the conditions under which telepathy occurs spontaneously, particularly in dreams. A team led by Stanley Krippner carried out a series of dream telepathy tests in which subjects slept in a soundproofed dream laboratory. Electrodes were attached to their heads to measure brain waves through an electroencephalograph (EEG), and eye movements were monitored. Rapid eye movements (REM) typically occur during dreams, and hence the researchers could tell when the subjects were dreaming. Before the subject went to bed, she met the sender, who thereafter stayed in another room, and in some cases in another building, miles away. When she was asleep and her eye movements indicated that she was dreaming, the sender opened a sealed package containing a randomly selected picture and concentrated on it, trying to influence the subject's dream. The subject was then woken by a buzzer and asked to describe her dream. Her comments were tape-recorded and transcribed. A panel of independent judges then compared the description of the dream with the pool of pictures from which the test picture had been selected at random, and decided which agreed best with the descriptions.

In some cases the agreement was very striking: one subject dreamed about buying tickets for a boxing match while the sender was looking at a picture of a boxing match. Sometimes the connection was more symbolic: for example, the subject dreamed of a dead rat in a cigar box, while the sender was looking at a picture of a dead gangster in a coffin.[12] In a total of 450 dream telepathy trials, the overall results were very significantly above the chance level.[13]

In the 1970s, several parapsychologists developed a new kind of telepathy test in which the subjects were in a state of mild sensory deprivation, because they thought that subjects were likely to do better when relaxed. The subject reclined in a comfortable chair in a special soundproofed room wearing headphones that played continuous white noise. Over their eyes were halved ping-pong balls, held on with tape, and they were bathed in dim red light. This set-up was called the *ganzfeld*, from a German word meaning 'whole field'. Meanwhile, the sender, in a separate soundproofed room, looked at a photograph or a video randomly selected from a pool of four. The subject spoke about her impressions and was recorded. At the end of the fifteen- or thirty-minute session, she was shown all four images from the pool in a random order and asked to rank them according to their closeness to her experience. If she placed the target image first, this counted as a hit.

By random guessing, subjects would have been right about one time in four, or 25 per cent of the time. By 1985, there had been twenty-eight *ganzfeld* studies in ten different laboratories, and the overall hit rate was 35 per cent, highly significant statistically. A well-known academic sceptic, Ray Hyman, agreed that the data showed a significant effect but thought it might have been due to a variety of flaws in the procedure. He and a leading researcher in the field, Charles Honorton, issued a joint communiqué specifying strict criteria that future tests should follow to eliminate possible flaws.[14]

Subsequent research followed these agreed criteria, and in this new series of studies the average hit rate was 34 per cent, again very significantly above the 25 per cent chance level.[15] In most of these studies the senders and receivers were strangers. When tests were done between people who knew each other well, for example mothers and daughters, the scores were far higher.[16]

Animal telepathy

Sir Rudolph Peters opened my mind to the possibility of telepathy and I am grateful to him for it. But as I began to take an interest in the subject, I soon realised that almost all psychic research and parapsychology concerned humans. Was this because psychic powers were special attributes of human beings? Or was it simply a reflection of the human-centred interests of researchers? Could animals be telepathic too? It seemed to me that if telepathy existed in humans, it might well be an ability shared with other animals.

At this stage, I came across a remarkable book written by William Long, called *How Animals Talk*, published in 1919.[17] Some of his most fascinating studies concerned wolves, which he observed for months on end in Canada. He found that separated members of wolf packs remained in contact with each other and responded to each other's activities over many miles. Wolves separated from the pack not only seemed to know what the others were doing but where they were. Their responsiveness involved more than following habitual paths, tracking scent trails, or by hearing howling or other sounds.

As Long pointed out, the same abilities may occur in domesticated animals. He was particularly interested in the ability of some dogs to know when their owners were coming home, and he described some simple experiments carried out by a friend of his with a return-anticipating dog. The dog started waiting for its owner soon after he had started his journey and waited for more than half an hour before he arrived home, even at non-routine times.

Unfortunately no one followed this lead. The subject of telepathy was taboo, and biologists avoided it. I started asking friends, family members and neighbours if they had ever noticed that their animals could anticipate when someone was coming home. I soon heard some very relevant stories. For example, in my home town, Newark-on-Trent, in Nottinghamshire, one of my neighbours was a widow who kept a cat that was very attached to her son, a seaman in the Merchant Navy. He did not tell her when he would be coming home on leave because he was afraid she would worry if he was late. But she knew anyway, because the cat sat on the front-door mat miaowing for an hour or two before he arrived. The cat's behaviour gave her time to get her son's room ready and prepare a meal for him.

I appealed through the media for people to tell me about their experiences with dogs and cats that anticipated arrivals, and soon received dozens of reports. By 2011, on my database I had more than a thousand accounts of dogs and more than six hundred of cats that anticipated their owners' returns. Many of these stories made it clear that the animals' responses were not simply reactions to the sounds of a familiar car, or familiar footsteps in the street. They happened too long in advance, and they also happened when people came home by bus or train. Nor was it just a matter of routine. Some people worked irregular hours, like plumbers, lawyers and taxi drivers, and yet those at home knew when they were coming because the dog or cat waited at the door or window, sometimes half an hour or more before the person arrived. More than twenty other species showed similar anticipatory behaviour, especially parrots and horses, but also a ferret, several bottle-fed lambs raised as pets, and pet geese. In random household telephone surveys in Britain and the United States, I found that in about 50 per cent of dog-owning households and about 30 per cent of cat-owning households, the animals were said to anticipate the arrival of a member of the family.[18]

I carried out experiments with return-anticipating dogs to find out whether they really did anticipate their owners' returns, even when they could not have known by 'normal' means. The first and most extensive tests were with a terrier called Jaytee, who lived near Manchester, England, with his owner, Pam Smart. Preliminary observations revealed that Jaytee usually started waiting for Pam just before she set off homewards, apparently when she formed the intention to do so. He did this whatever the time of day on 85 out of 100 return journeys. On some occasions when he failed to react he was sick; on others there was a bitch on heat in the next flat, showing that Jaytee could be distracted. But on 85 per cent of the occasions he seemed to anticipate Pam's return.[19]

In formal, randomised tests, Pam went at least five miles away from home. While she was out, the place where Jaytee waited was filmed continuously on time-coded videotape. She did not know in advance when she would be going home and she did so only when she received a message from me via a telephone pager at a randomly selected time. She travelled by taxi, in a different vehicle each time, to avoid any familiar car sounds. On average, Jaytee was at the window only four per cent of the time during the main period of Pam's absence, and 55 per cent of the time when she was on the way back. This difference was very significant statistically.[20]

I carried out many further filmed observations of Jaytee's behaviour,[21] and did similar experiments with other dogs, notably a Rhodesian Ridgeback called Kane.[22] Again and again, on film and under controlled conditions, these dogs anticipated their owners' returns.

Domestic animals seem to pick up their owners' thoughts and intentions in other ways: for example, many cats seem to know when they are going to be taken to the vet; they hide. Most dramatically, some animals seem to sense when their owners have had accidents or have died in distant places. On my database, there are 177 accounts of dogs apparently responding to the death or distress of their human companions, most by howling, whining or whimpering, and 62 accounts of cats showing signs of distress. Conversely, in 32 instances, people knew when their pet had died or was in dire need.[23]

The most remarkable animal I have come across is an African Grey parrot, N'kisi. His vocabulary of around 1,500 words is probably the largest ever recorded. When he was only about two years old, his owner, Aimée Morgana, noticed that he seemed to respond to her thoughts or intentions by saying what she was thinking. He slept in her bedroom, and woke her up on several occasions by commenting loudly on what she was dreaming.

She and I set up a controlled experiment in which she viewed a series of photographs in a random sequence while she was in a different room on a separate floor, being filmed continuously. The pictures were in a random order and represented twenty words in N'kisi's vocabulary, like 'flower', 'hug' and 'phone'. Meanwhile, N'kisi, who was alone, was also filmed continuously. He often said words that corresponded to the image she was viewing, and did so much more frequently than expected by chance. The results were highly significant statistically.[24]

The natural history of human telepathy

The laboratory investigations of telepathy by parapsychologists provide evidence that it happens, but shed little light on telepathy in real-life situations.

When Laurens van der Post was living with Bushmen in the Kalahari desert, in southern Africa, he found that they seemed to be in telepathic contact on a regular basis. On one occasion he went hunting with a group of men and they killed an eland about fifty miles away from the camp. As they were driving back in a Land Rover laden with meat, van der Post asked one of the Bushmen how the people would react when they learned of their success. He replied, 'They already know. They know by wire . . . We Bushmen have a wire here' – he tapped his chest – 'that brings us news.' He was comparing their method of communication with the white man's telegram or 'wire'. Sure enough, when they approached the camp, the people were singing the 'Eland Song' and preparing to give the hunters the greatest of welcomes.[25]

In most, if not all, traditional societies, telepathy seems to be taken for granted and put to practical use. Many travellers in Africa reported that people seem to know when others to whom they were attached were coming home. The same occurred in rural Norway, where there is a special word for the anticipation of arrivals: *vardøger*. Typically, someone at home heard a person approaching the house and coming in, yet nobody physically did so. Soon afterwards the person actually arrived. Similarly, the 'second sight' of some inhabitants of the Scottish Highlands included visions of arrivals before the person in the vision appeared.

Unfortunately, most anthropologists who have lived with traditional peoples have not studied these aspects of their behaviour, or at least they have not reported them. Materialist taboos have inhibited their spirit of enquiry. As a result, very little is known about the natural history of telepathy and other psychic phenomena in other cultures.

In an attempt to find out more about telepathy in modern

societies, I launched a series of appeals for information through the media in Europe, North America and Australia. Over a period of fifteen years, I built up a database of human experiences, similar to my database on unexplained powers of animals, with more than four thousand case histories, classified into more than sixty categories.

Many cases of apparent telepathy occurred in response to other people's needs. For example, hundreds of mothers told me that when they were still breastfeeding, they knew when their baby needed them, even from miles away. They felt their milk let down. (The milk let-down reflex is mediated by the hormone oxytocin, sometimes called the love hormone, and is normally triggered by hearing the baby cry. The nipples start leaking milk and many women feel a tingling sensation in their breasts.) When mothers who were away from their babies felt their milk let down, most of them took it for granted that their baby needed them. They were usually right. They did not experience their milk letting down because they started thinking about the baby; they started thinking about their baby because their milk let down for no apparent reason. Their response was physiological.

With the help of a midwife, I carried out a study of nine nursing mothers in North London over a two-month period. While the mothers were away from their babies, all the times at which their milk let down were recorded; meanwhile the baby-sitters noted when the babies showed signs of distress. After eliminating events that could have been due to synchronised rhythms in mother and baby at regular feeding times, most of the unexpected milk let-downs did indeed coincide with the babies' distress. The statistical odds against their occurring by chance were a billion to one. In other words, it was highly unlikely that the mothers' responses were random coincidences.

A telepathic connection between mothers and their babies makes good sense in evolutionary terms. Mothers who could tell at a distance when their babies needed them would tend to have babies that survived better than the babies of insensitive mothers.

Telepathic connections between mothers and their children

often seem to continue even when the children have grown up. Many stories on my database concern mothers who went to or telephoned their children when they could not have known by any conventional means that they were in distress.

Until the invention of modern telecommunications, telepathy was the only way in which people could be in touch at a distance instantly. In most respects telepathy has now been superseded by the telephone – but it has not gone away. Telepathy now occurs most commonly in connection with telephone calls.

Telephone telepathy

The commonest kind of story about apparent telepathy concerned telephone calls. Hundreds of people told me that they had thought of someone for no apparent reason, and then that person rang in a way that seemed uncanny. Or they knew who was calling when the phone rang before they looked at a caller ID display or answered. I followed up these stories with a series of surveys in Europe and in North and South America. On average, 92 per cent of the respondents said they had thought of someone as the telephone rang, or just before, in a way that seemed telepathic.[26]

When I talked to friends and colleagues about this phenomenon, most agreed it seems to happen. Some people simply accepted that it was telepathic or intuitive; others tried to explain it 'normally'. Almost everyone came up with one or both of these two arguments. First, they said, you think about other people frequently; then sometimes, by chance, somebody rings while you are thinking about them; you imagine it is telepathy, but you forget the thousands of times you were wrong. The second argument was that when you know someone well, your familiarity with his routines and activities enables you to know when he is likely to ring, even though this knowledge may be unconscious.

I searched the scientific literature to find out if these standard arguments were supported by any data or observations. I could find no research whatever on the subject. The standard sceptical arguments were evidence-free speculations. In science it is not enough to put forward a hypothesis: it needs to be tested.

I designed a simple procedure to test both the chance-coincidence theory and the unconscious-knowledge theory experimentally. I recruited subjects who said they quite frequently knew who was calling before answering the phone. I asked them for the names and telephone numbers of four people they knew well, friends or family members. The subjects were then filmed continuously throughout the period of the experiment alone in a room with a landline telephone, without a caller ID system. If there was a computer in the room, it was switched off, and the subjects had no mobile phone. My research assistant or I selected one of the four callers at random by the throw of a die. We rang up the selected person and asked him to phone the subject in the next couple of minutes. He did so. The subject's phone rang, and before answering it she had to say to the camera who, out of the four possible callers, she felt was on the line. She could not have known through knowledge of the caller's habits and daily routines because, in this experiment, the callers rang at times randomly selected by the experimenter.

By guessing at random, subjects would have been right about one time in four, or 25 per cent. In fact, the average hit rate was 45 per cent, very significantly above the chance level. None of the subjects was right every time, but they were right much more than they should have been if the chance coincidence theory were true. This above-chance effect has been replicated independently in telephone telepathy tests at the universities of Freiburg, Germany, and Amsterdam, Holland.[27]

In some of our tests, there were two familiar callers and two unfamiliar callers whom the subjects had never met but whose names they knew. The hit rate with unfamiliar callers was near the chance level; with the familiar callers it was 52 per cent, about twice the chance level. This experiment supported the idea that telepathy occurs more between people who are bonded to each other than between strangers.

For some of our experiments we recruited young Australians,

New Zealanders and South Africans living in London. Some of their callers were back home, thousands of miles away, and the others were new acquaintances in England. In these tests, the hit rates were higher with their nearest and dearest far away than with people in Britain they had met only recently, showing that emotional closeness is more important than physical proximity.[28]

Other researchers have also found that telepathy does not seem to be distance-dependent.[29] At first sight this seems surprising, because most physical influences, like gravitation and light, fall off with distance. But the physical phenomenon most analogous to telepathy is quantum entanglement, also known as quantum non-locality, which does not fall off with distance.[30] When two quantum particles have been part of the same system and move apart, they remain interconnected or entangled in such a way that a change in one is associated with an immediate change in the other. Albert Einstein famously described this effect as 'spooky action at a distance'.[31]

Telepathy has continued to evolve along with modern technologies. Many people say they have the experience of thinking of someone who then emails them, or sends them a text message. Experiments with emails and text messages using similar methods to the telephone tests have also given positive, highly significant results.[32] As in telephone tests, the effect occurred more with familiar people and it did not fall off with distance. The same was true of automated telepathy tests on the Internet.[33]

I do not know to what extent people can learn to be more sensitive telepathically, but several automated tests are now available, including a test that runs on mobile phones, enabling those who are interested to find out for themselves.[34]

Telepathy involves picking up feelings, needs or thoughts at a distance, across space. Some other phenomena are also spatial, like the sense of being stared at and remote viewing. By contrast, premonitions, precognition and presentiments are related to future events, and imply links across time from the future to the present.

Animal premonitions of disasters

Premonition means warning in advance; precognition means knowing in advance; and presentiment means feeling in advance.

There are many examples of animals seeming to feel when a disaster is about to strike. Ever since classical times, people have reported unusual animal behaviour before earthquakes, and I have collected a large body of evidence for unusual animal behaviour before recent earthquakes, including the 1987 and 1994 earthquakes in California; 1995 in Kobe, Japan; 1997 near Assisi, Italy; 1999 in Izmit, Turkey; and 2001 near Seattle, Washington. In all these cases there were many reports of wild and domesticated animals behaving in fearful, anxious or unusual ways several hours or even days before the earthquakes. Dogs were howling for hours beforehand, and many cats and birds were behaving unusually.[35]

One of the very few systematic observations of animal behaviour before, during and after an earthquake concerns toads in Italy. A British biologist, Rachel Grant, was carrying out a study of mating behaviour in toads for her PhD project at San Ruffino Lake in central Italy in the spring of 2009. To her surprise, soon after the beginning of the mating season in late March, the number of male toads in the breeding group suddenly fell. From more than ninety being active on 30 March, there were almost none on 31 March and in early April. As Grant and her colleague Tim Halliday observed, 'This is highly unusual behaviour for toads; once toads have appeared to breed, they usually remain active in large numbers at the breeding site until spawning has finished.' On 6 April, Italy was shaken by a 6.4-magnitude earthquake, followed by a series of aftershocks. The toads did not resume their normal breeding behaviour for another ten days, two days after the last aftershock. Grant and Halliday looked in detail at the weather records for this period but found nothing unusual. They were forced to the conclusion that the toads were somehow detecting the impending earthquake some six days in advance.[36]

No one knows how some animals sense when earthquakes are imminent. Perhaps they pick up subtle sounds or vibrations in the earth. But if animals can predict earthquake-related disasters by sensing slight tremors, why can't seismologists do so? Or maybe they respond to subterranean gases released prior to earthquakes, or react to changes in the earth's electrical field. But they may also sense in advance what is about to happen in a way that lies beyond current scientific understanding, through some kind of presentiment.

Similarly, many animals seemed to anticipate the great Asian tsunami on 26 December 2004, although their reactions were much closer to the actual event. Elephants in Sri Lanka and Sumatra moved to high ground before the giant waves struck; they did the same in Thailand, trumpeting beforehand. According to villagers in Bang Koey, Thailand, a herd of buffalo were grazing by the beach when they 'suddenly lifted their heads and looked out to sea, ears standing upright'. They turned and stampeded up the hill, followed by bewildered villagers, whose lives were thereby saved. At Ao Sane beach, near Phuket, dogs ran up to the hill tops, and at Galle in Sri Lanka, dog owners were puzzled when their animals refused to go for their usual morning walk on the beach. In Cuddalore District in south India, buffaloes, goats and dogs escaped by moving to higher ground, and so did a nesting colony of flamingos. In the Andaman Islands, 'stone age' tribal groups moved away from the coast before the disaster, alerted by the behaviour of animals.[37]

How did they know? The usual speculation is that the animals picked up tremors caused by the under-sea earthquake. But this explanation is unconvincing. There would have been tremors all over South East Asia, not just in the afflicted coastal areas.

Some animals anticipate other kinds of natural disaster like avalanches,[38] and even man-made catastrophes. During the Second World War, many families in Britain and Germany relied on their pets' behaviour to warn them of impending air raids before official warnings were given. The animal reactions occurred when enemy planes were still hundreds of miles away, long before the animals could have heard them coming. Some dogs in London anticipated the explosion of German V-2 rockets.

These missiles were supersonic and could not have been heard in advance.[39]

With very few exceptions, the ability of animals to anticipate disasters has been ignored by Western scientists; the subject is taboo. By contrast, since the 1970s, in earthquake-prone areas of China, the authorities have encouraged people to report unusual animal behaviour, and Chinese scientists have an impressive track record in predicting earthquakes. In several cases they have issued warnings that enabled cities to be evacuated hours before devastating earthquakes struck, saving tens of thousands of lives.[40]

By paying attention to unusual animal behaviour, as the Chinese do, earthquake and tsunami warning systems might be feasible in parts of the world that are at risk from these disasters. Millions of people could be asked to take part in this project through the media. They could be told what kinds of behaviour their pets and other animals might show if a disaster were imminent – in general, signs of anxiety or fear. If people noticed these signs, or any other unusual behaviour, they would immediately telephone a hotline with a memorable number – for example, in California, 1-800-PET QUAKE. Or they could send a message on the Internet.

A computer system would analyse the places of origin of the incoming messages. If there was an unusually large number, it would signal an alarm, and display on a map the places from which the calls were coming. There would probably be a background of false alarms from people whose pets were sick, for example, and there might also be scattered hoax calls. But if there was a sudden surge of calls from a particular region, this could indicate that an earthquake or tsunami was imminent.

Exploring the potential for animal-based warning systems would cost relatively little. From a practical point of view, it does not matter how animals know: they can give useful warnings whatever the explanation. If it turns out that they are indeed reacting to subtle physical changes, then seismologists should be able to use instruments to make better predictions themselves. If it turns out that presentiment plays a part, we will learn something important about the nature of time and causation. By ignoring animal premonitions, or by explaining them away, we will learn nothing.

Human premonitions and precognitions

Sixteen-year-old Carole Davies was about to leave a games arcade in London with some friends when it began to rain heavily. The entrance became crowded as people came in from the street to shelter. Carole said:

> While standing there looking out into the night, I had a sense of danger. Then I saw what looked like a picture in front of me showing people on the floor and with tiles and metal girders on them. I looked round and up and realised this was to happen here. I began to shout at people to get out. No one listened. I ran through the rain, with my friends following, to a nearby café. After a while we heard sirens that stopped outside the arcade building. We all ran down the road to find out what had happened. It was just as I had seen. A man I had shouted at was being pulled from under the debris.

During wars, people tend to be more alert to danger, and indeed there is more danger. For example, Charles Bernuth, serving in the US Seventh Army in the Second World War, took part in the invasion of Germany. Soon after crossing the Rhine, he was driving along an autobahn at night with two fellow officers:

> All of a sudden, I got the still small voice. Something was wrong with the road. I just knew it. I stopped, amid the groans and jeers of the other two. I started walking along the road. About fifty yards from where I had left the jeep, I found out what was wrong. We were about to go over a bridge – only the bridge wasn't there. It had been blown up and there was a sheer drop of about seventy-five feet.

People who had these premonitions survived because they heeded the sense of danger.

On my database there are 842 cases of human premonitions, precognitions, or presentiments. Of these, 70 per cent are about dangers, disasters, or deaths; 25 per cent are about neutral events; and only five per cent are of happy events, like meeting a future spouse, or winning a raffle. Dangers, deaths and catastrophes predominate. This agrees with a survey of well-authenticated cases of precognition collected by the Society for Psychical Research in which 60 per cent concerned deaths or accidents. Very few were of happy events. Most of the others were trivial or neutral, although some were very unusual.[41] In one such case, the wife of the Bishop of Hereford dreamed that she was reading the morning prayers in the hall of the Bishop's Palace. After doing so, on entering the dining room, she saw an enormous pig standing beside the table. This dream amused her, and she told it to her children and their governess. She then went into the dining room and an escaped pig was standing in the exact spot where she had seen it in her dream.[42]

Many precognitions occur in dreams, although often only the most dramatic or bizarre are remembered. In the early twentieth century, a British aeronautical engineer, J. W. Dunne, made an astonishing discovery, summarised in his remarkable book *An Experiment With Time*.[43] He found he often dreamed about events that were about to happen, but usually forgot these dreams. Only by keeping careful records of his dreams, writing them down as soon as he awoke, did the phenomenon become clear. He also found that he sometimes had experiences that seemed familiar – sometimes called *déjà vu*, French for 'already seen' – and by looking back over his records found that they corresponded to recent dreams he had forgotten.

Subsequent studies have confirmed Dunne's observations. Parapsychologists have also found statistical evidence for precognition in laboratory tests. Although the effects in these very artificial experiments were usually small, taken together they were very significant statistically.[44]

Presentiments

A presentiment is a *feeling* that something is about to happen, but without any conscious awareness of what it is. Some of the most innovative research in modern parapsychology has shown that presentiments can be detected physiologically.

In the mid-1990s, in the United States, Dean Radin and his colleagues devised an experiment to test for presentiment in which a subject's emotional arousal was monitored automatically by measuring changes in skin resistance with electrodes attached to the fingers, as in a lie-detector test. As people's emotional states change, so does the activity of the sweat glands, resulting in changes in electrodermal activity, registered in a computerised recording device.

In the laboratory it is relatively easy to produce measurable emotional changes in subjects by exposing them to noxious smells, mild electric shocks, emotive words, or provocative photographs. Radin's experiments used photographs. Most were pictures of emotionally calm subjects like landscapes, but some were shocking, like pictures of corpses that had been cut open for autopsies; others were pornographic. A large pool of these 'calm' and 'emotional' images was stored within the computer.

In Radin's experiments, when calm pictures appeared on the screen, the subjects remained calm, and when emotional images were displayed, they were emotionally aroused, as shown by an increase in electrodermal activity. This is not surprising. But when emotional images were *about to* appear, the increase in electrodermal activity began *before* the picture appeared on the screen, three or four seconds in advance. The picture that appeared on the screen was selected at random by the computer only a millisecond in advance. No one, not even the experimenter, knew what it would be when the subjects began to react.[45] Other researchers have found similar results.[46]

One of the most interesting findings of precognition and presentiment research is that people seem to be influenced by *themselves* in the future, rather than by objective events. Precognitions are like memories of the future. Presentiments seem to involve a physiological back-flowing from future states of alarm or arousal, a flow of causation moving in the opposite direction to energetic causation. This is in agreement with the way that attractors pull organisms towards their inherited or learned goals, with flows of influence from virtual futures through the present towards the past (see Chapter 5). It also agrees with Alfred NorthWhitehead's idea of minds working from the future (see Chapter 4).

What skeptics say

'Organized skeptics' in Britain use the American spelling, with a 'k', rather than the usual English spelling, with a 'c'.

Informed skeptics do not deny that there is a lot of experimental evidence that suggests psychic phenomena are real, but they point out that no experiment is perfect, the evidence is not 100 per cent positive, and that for such an unlikely proposition, vastly more evidence is needed than in more orthodox science.[47] Skeptics feel free to move the scientific goalposts as far as they like. There is not enough evidence yet, they say, and for some skeptics there never will be.[48]

Skeptical organisations are the principal defenders of the belief that the psychic phenomena are illusory: they seek to debunk or deny any evidence that suggests they might be wrong. The best established of these groups is the US Committee for Skeptical Inquiry (CSI), which used to be called CSICOP (pronounced sigh-cop), the Committee for the Scientific Investigation of Claims of the Paranormal. CSI's publication, *The Skeptical Inquirer*, has a circulation of about 50,000. Members of skeptical organisations often think of themselves as lonely defenders of science and reason against the forces of superstition and credulity; they see their debunking activities as 'battles' against the insidious forces of irrationalism. Their opponents see them as self-appointed vigilantes.[49]

These effects of these well-organised and well-funded skeptical campaigns are not simply intellectual, but political and economic. Through maintaining the taboo against the 'paranormal', they ensure that most universities avoid this controversial area altogether, despite great public interest in the subject. Their main emphasis is on counteracting 'claims of the paranormal' in the serious media, either by attacking journalists or publications that report any positive evidence or by insisting that a skeptic is given the opportunity to deny that the evidence has any scientific validity.[50]

I have had many encounters with skeptics, described in detail elsewhere.[51] In almost every case, they were not only ignorant of the evidence, but uninterested in it. Here are three examples.

In 2004, I took part in a debate on telepathy with Lewis Wolpert at the Royal Society of Arts in London, with an eminent lawyer in the chair. Wolpert was a professor of biology at University College, London, and a former chairman of the British Committee on the Public Understanding of Science. For years he had been a faithful standby for journalists as a denouncer of the paranormal, always ready to provide a sceptical comment. We were each given thirty minutes to present our cases. Wolpert spoke first. He said that research on telepathy was 'pathological science', but after saying, 'The whole issue is about evidence,' he presented none. He simply stated, 'There is zero evidence to support the idea that thoughts can be transmitted from a person to an animal, from an animal to a person, from a person to a person, or from an animal to an animal.' He used up only half his allotted time.

I summarised evidence for telepathy from thousands of scientific trials and showed a video of recent experiments. Wolpert was sitting on the stage in front of the screen, staring ahead, tapping a pencil on the table, sighing as if bored. He did not turn around to see the evidence behind him. According to a report on the debate in *Nature*, 'few members of the audience seemed to be swayed by his [Wolpert's] arguments ... Many in the audience variously accused Wolpert of "not knowing the evidence" and being "unscientific".'[52] For anyone who wants to hear both sides for themselves, the debate is online in streaming audio.[53]

Second, I was invited to speak at the 12th European Skeptics Congress in Brussels, Belgium, in 2005. I took part in a debate on telepathy with Jan Nienhuys, the secretary of a Dutch skeptical organisation, Stichting Skepsis. I presented evidence for telepathy, reviewing research by others and by myself. Nienhuys responded by arguing that telepathy was impossible on theoretical grounds and therefore all the evidence was flawed. He said that the more statistically significant my experimental results were, the greater the errors must be. I asked him to specify these errors, but he could not. He admitted he had not actually read my papers or looked at the evidence. In an account of the debate, an independent observer, Richard Hardwick, a scientist at the European Commission, wrote, 'It seems Dr Nienhuys had not done his homework. He did not have any data or analyses to hand, and his attack fizzled out.'[54]

In 2006, the British TV company Channel 4 broadcast a two-part diatribe by Richard Dawkins against religion, called *The Root of All Evil?* Soon afterwards, the same production company, IWC Media, told me that Dawkins wanted to visit me to discuss my research on unexplained abilities of people and animals for a new TV series. I was reluctant to take part because I expected that it would be as one-sided as Dawkins's previous series. But the company's representative, Rebecca Frankel, assured me that they were now more open-minded. She told me, 'We are very keen for it to be a discussion between two scientists, about scientific modes of enquiry.' On the understanding that Dawkins was interested in discussing evidence, and with the written assurance that the material would be edited fairly, I agreed to meet him and we fixed a date. I was still not sure what to expect. Was he going to be dogmatic, with a mental firewall that blocked out any evidence that went against his beliefs? Or would he be fun to talk to?

Dawkins duly came to call. The director, Russell Barnes, asked us to stand facing each other; we were filmed with a hand-held camera. Dawkins began by saying that he thought we probably agreed about many things, 'But what worries me about you is that you are prepared to believe almost anything. Science should be based on the minimum number of beliefs.'

I agreed that we had a lot in common, 'But what worries me

about you is that you come across as dogmatic, giving people a bad impression of science, and putting them off.'

Dawkins then said that in a romantic spirit he himself would like to believe in telepathy, but there just wasn't any evidence for it. He dismissed all research on the subject out of hand, without going into any details. He said that if it really occurred, it would 'turn the laws of physics upside down', and added, 'Extraordinary claims require extraordinary evidence.'

'This depends on what you regard as extraordinary', I replied. 'Most people say they have experienced telepathy, especially in connection with telephone calls. In that sense, telepathy is ordinary. The claim that most people are deluded about their own experience is extraordinary. Where is the extraordinary evidence for that?' He could not produce any evidence at all, apart from generic arguments about the fallibility of human judgement. He took it for granted that people want to believe in 'the paranormal' because of wishful thinking.

We then agreed that controlled experiments were necessary. I said that this was why I had actually been doing such experiments, including tests to find out if people really could tell who was calling them on the telephone when the caller was selected at random. The previous week, I had sent Dawkins copies of some of my papers in scientific journals so that he could examine the data before we met. I suggested that we actually discuss the evidence. He looked uneasy and said, 'I don't want to discuss evidence.'

'Why not?' I asked.

He replied, 'There isn't time. It's too complicated. And that's not what this programme is about.' The camera stopped.

The director confirmed that he, too, was not interested in evidence. The film he was making was another Dawkins polemic against irrational beliefs. I said, 'If you're treating telepathy as an irrational belief, surely evidence about whether it exists or not is essential for the discussion. If telepathy occurs, it's not irrational to believe in it. I thought that's what we were going to talk about. I made it clear from the outset that I wasn't interested in taking part in another low-grade debunking exercise.'

Dawkins replied, 'It's not a low-grade debunking exercise; it's a high-grade debunking exercise.'

I said I had been assured that this was to be a balanced scientific discussion about evidence. Russell Barnes asked to see the emails I had received from his assistant. He read them with obvious dismay, and said the assurances she had given me were wrong. In that case, I said, they were visiting me under false pretences. The team packed up and left. The series, broadcast in 2007, was called *Enemies of Reason*.

Richard Dawkins has long proclaimed, 'The paranormal is bunk. Those who try to sell it to us are fakes and charlatans.' *Enemies of Reason* was intended to popularise this belief. But does his crusade really promote the Public Understanding of Science, of which he was professor at Oxford? Should science be a fundamentalist belief-system? Or should it be based on open-minded enquiry into the unknown?

In no other field of scientific endeavour do otherwise intelligent people feel free to make public claims based on prejudice and ignorance. No one would denounce research in physical chemistry, say, while knowing nothing about the subject. Yet in relation to psychic phenomena, committed materialists feel free to disregard the evidence and behave irrationally and unscientifically while claiming to speak in the name of science and reason. They abuse the authority of science and bring rationalism into disrepute.

What difference does it make?

Dropping the taboo against psychic phenomena would have a liberating effect on science. Scientists would no longer feel the need to pretend these phenomena are impossible. The word 'skepticism' would be liberated from its association with dogmatic denial. People would feel free to talk openly about their own experiences. Open-minded research would be able to take place within universities and scientific institutions, and some of this research could be applied in useful ways, for example in the development of animal-based warning systems for earthquakes and tsunamis. Public funding for psychic research and parapsychology could reflect the widespread interest in these areas of research, and increase public interest in science. The educational system would be free to teach students about psychic research instead of ignoring or dismissing it. Anthropologists would be liberated from the taboo that stops them studying psychic skills that may be better developed in traditional societies than in our own. Above all, research on these phenomena would contribute to a larger and more inclusive understanding of the nature of minds, social bonds, time and causation.

Questions for materialists

If you think telepathy and precognition are theoretically impossible, or very improbable, can you explain why?

Have you ever looked at the evidence for psychic phenomena? If so, can you summarise it, and explain what is wrong with it?

Have you yourself ever had a seemingly telepathic experience?

What might persuade you to change your mind?

SUMMARY

Most people claim to have had telepathic experiences. Numerous statistical experiments have shown that information can be transmitted from person to person in a way that cannot be explained in terms of the normal senses. Telepathy typically happens between people who are closely bonded, like mothers and children, spouses and close friends. Many nursing mothers seem to be able to detect when their babies are in distress when they are miles away. The commonest kind of telepathy in the modern world occurs in connection with telephone calls when people think of someone who then rings, or just know who's calling. Numerous experimental tests have shown that this is a real phenomenon. It does not fall off with distance. Social animals seem to be able to keep in touch with members of their group at a distance telepathically, and domesticated animals, like dogs, cats, horses and parrots, often pick up their owners' emotions and intentions at a distance, as shown in experiments with dogs and parrots. Other psychic abilities include premonitions and precognitions, as shown by animals' anticipation of earthquakes, tsunamis and other disasters. Human premonitions usually occur in dreams or through intuitions. In experimental research on human presentiments, future emotional events seem able to work 'backwards' in time to produce detectable physiological effects.

10

Is Mechanistic Medicine the Only Kind that Really Works?

Modern medicine is amazingly successful. Its achievements would have seemed miraculous a hundred years ago. Heart transplants, keyhole surgery, hip replacements and *in-vitro* fertilisation are just a few of the interventions that have changed millions of people's lives. Together with immunisation programmes and advances in public health, 'miracle drugs', like antibiotics, have affected all humanity, reducing infant mortality and increasing life expectancy almost everywhere.

There is no doubt that modern medicine works very well. Yet it has important limitations that are becoming increasingly apparent. The great advances of medicine over the last century are running out of steam. The rate of discovery is slowing, despite ever-increasing investment in research. There is a dearth of new drugs, and medicine is becoming prohibitively expensive.

The mechanistic approach is at its best when dealing with mechanical aspects of the body, like defective joints, decayed teeth, faulty heart valves and blocked arteries, or infections curable with antibiotics. But it suffers from tunnel vision: all living organisms, including people, are physico-chemical machines, or 'lumbering robots'. Hence a materialist medical system confines its attention to the physical and chemical aspects of human beings, treating them through surgery and drugs, while ignoring anything that does not fit in.

Doctors are often intensely annoyed by the existence of rival medical systems, including homeopathy, chiropractic and traditional Chinese medicine, all of which claim to be able to heal the sick. From the point of view of a committed materialist, none of these other systems can really work. Either the patients would have got better anyway, or else the benefits of alternative and complementary therapies are simply a product of the placebo effect.

The belief that only mechanistic medicine is truly effective has enormous political and economic consequences. In most countries, government funding for medical research, totalling many billions of dollars, is exclusively confined to mechanistic medicine. Most national health services and medical insurance companies are equally mechanistic in their approach.

In this chapter I discuss the strengths and limitations of mechanistic medicine. Its strengths depend on working with the natural capacities for healing and resistance to disease that are inherent in all people, and indeed in all forms of life. Its mechanistic emphasis results in spectacular chemical and physical cures through drugs and surgery. But the failure to recognise the power of minds means that it is weakest when dealing with the healing effects of beliefs, expectations, social relationships and religious faith. Yet through medical research itself, the importance of belief has been revealed again and again by placebo responses. I end by exploring the possibility of a more inclusive approach to health and healing.

I adopt a historical approach because I believe it offers the best way of understanding our present situation. It is particularly important in revealing which aspects of medicine depend on the materialist worldview, and which result from pragmatic discoveries that are not dependent on any particular philosophy of nature.

Natural capacities to heal and resist disease

In order to see any medical system in perspective, it is important to remember that animals and plants have been regenerating after damage, healing themselves and defending themselves against infections throughout the entire history of life on earth. All of us are descended from animal and human forebears that survived and reproduced for hundreds of millions of years before the advent of doctors. We would not be here if it were not for our ancestors' innate capacities to heal and resist diseases. Medicine can help and enhance these capacities, but it builds on foundations that have evolved over vast aeons of time, continually subject to natural selection.

The ability to heal after wounding and to regenerate after damage is common to almost all forms of life. If plants are injured or attacked by diseases, they often seal off the wounded area and compensate by growing new tissues. Small parts of plants can regenerate into entire new organisms: cuttings taken from a willow tree can grow into new willow trees. Similarly, many animals have astonishing powers of regeneration. A flatworm can be cut into small pieces, and each can grow into a whole new worm. The leg can be cut off a newt and it will grow a new leg; the lens can be removed from its eye and it will form a new lens from the edge of its iris.[1] Even in humans, the skin regenerates after wounding, so does the liver, and the intestinal lining and blood cells are continually replaced (see Chapter 5).

Many organisms have the power to resist diseases through an immune response. In bacteria, enzyme systems attack invading viruses; in plants, immune systems recognise and respond to pathogens by producing chemicals that kill them or inhibit their growth.[2] Likewise, in invertebrate animals, such as insects, immune systems attack and destroy invading organisms. The immune systems in vertebrates go further and remember specific pathogens, mounting a stronger attack the next time they encounter the invading organism.

People have known for a long time that exposure to a disease can confer immunity later to the same disease. In his description of the plague in Athens in 430 BC, the classical Greek historian Thucydides was the first to note that people who had recovered from the plague were able to nurse the sick without catching it.[3] Building on observations like this, at least six hundred years ago, some Arabs and Chinese were inoculating people with material taken from the pustules of others suffering from a mild form of smallpox. As a result, most of those who were inoculated contracted a mild form of the disease and survived unscathed when exposed to the disease again.[4]

In 1718, the wife of the British ambassador in Istanbul, Lady Mary Wortley Montagu, inoculated her own children in this way. She had lost her brother to smallpox, and the disease had marred her own beauty a few years earlier, so she took a keen interest in this procedure when she learned about it from Turkish women. On her return to England she promoted the method enthusiastically. Several members of the Royal Family were inoculated, and the practice became widespread, although about three per cent of those who were inoculated died of the disease.

In the 1790s, Edward Jenner, an English physician, modified the technique of immunisation in the light of the observation that dairymaids did not catch smallpox after they had been infected with cowpox, a much milder disease. Jenner developed the technique of vaccination (from the Latin *vacca*, cow) by taking fluid from pustules on infected dairymaids and deliberately infected children through a small wound in the skin. In 1853 a law required universal vaccination against smallpox in England and Wales, and vaccinations and inoculations for other diseases were developed and widely applied in the twentieth century, long before the immune system was described in cellular and molecular detail.

The World Health Organisation declared in 1979 that smallpox had been eradicated entirely.

Hygiene and public health

In the early nineteenth century, a number of epidemiologists and clinicians came to the conclusion that diseases like puerperal fever and cholera were spread through microscopic germs and could be counteracted by improvements in hygiene and sanitation. Louis Pasteur's discovery of specific germs in the 1860s and the subsequent development of the germ theory of disease underlay a series of preventive measures and public-health policies that resulted in a dramatic decline in deaths from epidemics.

The improvements in public health through the prevention of infectious diseases were triumphs of scientific research, water and sewage engineering, public policy initiatives and health education. Neither the discovery of germs as the infectious agents of disease nor the building up of immunity through vaccination depended on any particular dogma. None of the advances was specifically dependent on the mechanistic theory of life or the materialist worldview.

Much of the success of medicine in the twentieth century was due to the prevention of diseases through immunisation and improved hygiene. As these preventive measures were spread worldwide, there was a global reduction in infant and child mortality and there were far fewer epidemics. One result was a dramatic increase in the world's population from 1 billion in 1800, to 3 billion in 1960, and 7 billion in 2012.

Cures for infections

One of the most iconic discoveries of twentieth-century medicine was penicillin, found by accident by a microbiologist, Alexander Fleming, in 1928. When he was culturing bacteria in Petri dishes, one of the dishes was contaminated with a mould, *Penicillium notatum*. Fleming noticed that all around the mould the bacteria were dying. He found that juice extracted from the mould, which he called penicillin, could inhibit the growth of a wide range of other bacteria. But he assumed that penicillin would be too toxic to be of any medical use, and he did not pursue it.

Fleming's work was rediscovered ten years later by Howard Florey and Ernst Chain in Oxford, and it was not until 1941 that its full potential became apparent. It was a wonder drug that rapidly produced spectacular effects. Not only did it cure acute potentially lethal infections like septicaemia, pneumonia and meningitis, but also chronic infections of sinuses, joints and bones. With the other antibiotics that followed in its wake, it transformed both the public and doctors' perception of what medicine could do.[5] Yet scientists did not *invent* antibiotics; *Penicillium notatum* and other organisms produced them for their own purposes. Antibiotics were a gift of nature.

Together with improvements in hygiene and programmes of mass immunisation, the discovery of antibiotics meant that the death rate from infectious diseases plummeted. Dreaded diseases like cholera, typhoid, tuberculosis and polio no longer killed people by the million. These staggering advances changed the very conditions of human life.

In the late twentieth century, the power of antibiotics was extended yet further by the surprising discovery that stomach ulcers, previously thought to be caused by excess acidity in the stomach or by stress, were in fact the result of infection with a hitherto unknown bacterium, *Helicobacter pylori*, and could be cured with antibiotics.[6]

The young were the major beneficiaries of the control of infectious diseases with immunisation and antibiotics. Infant and child mortality, once common, are now rare. Today the most serious medical problems among young people are inherited diseases, like cystic fibrosis, allergies, like asthma, or accidents. The main challenges that now face medicine are the diseases of old age, including cancer and diseases of the circulatory system, as well as chronic 'degenerative' diseases, like arthritis and dementia. On the whole, most adults are remarkably healthy until their fifties or sixties. Nevertheless, there are still several major diseases that afflict the middle-aged, including diabetes, rheumatoid arthritis, multiple sclerosis, Parkinson's disease and schizophrenia. The causes of most of these diseases are unknown.[7]

Meanwhile, the germs that cause infectious diseases continue to evolve. The appearance of new diseases like AIDS and the evolution of bacteria resistant to antibiotics still pose major problems.

New drugs

Throughout human history, people all over the world have used plants in herbal remedies, but it was not until the nineteenth century that chemists began to isolate 'active principles' from medicinal plants: morphine from opium poppies, cocaine from coca leaves, nicotine from tobacco, quinine from cinchona, salicylic acid from willow bark, and a host of other pharmacologically active compounds.[8] These purified drugs were more reliable and predictable in their effects than variable herbal materials. And once the pure drugs had been chemically identified, they could also be modified chemically to produce new substances that were more powerful or had fewer side effects than the naturally occurring compounds, such as acetyl salicylic acid (aspirin) from salicylic acid and diacetyl morphine (heroin) from morphine. In some cases a range of compounds with similar structures was created, known as analogues: for example, xylocaine, amylocaine and procaine, analogues of cocaine, widely used as local anaesthetics.

The discovery of penicillin and other antibiotics took this process further, and their spectacular success gave a boost to the quest for new drugs. If these naturally occurring non-toxic chemicals could cure appalling diseases and make all the difference between life and death, then why should not other illnesses yield to simple chemical solutions? Could chemical cures for cancer or schizophrenia be waiting to be discovered?

Like the drugs derived from medicinal plants, the antibiotics were gifts of nature, but their identification, purification and modification depended on the art of chemistry. The isolation of drugs from plants, fungi and bacteria continued at an increasing pace, and chemical compounds derived from natural sources, together with synthetic variants on them, account for about 70 per cent of the drugs in modern medicine.[9]

The other principal method of drug discovery is by trial and error. Drug companies test large numbers of chemicals isolated from plants or synthesised by chemists to find out if any of them have useful effects, while being sufficiently non-toxic. This process, called screening, is usually carried out on animals, although some tests now use animal or human cells grown *in vitro*, literally meaning 'in glass'. Since the 1950s, drug companies have screened tens of thousands of compounds and have discovered several important new drugs, including taxol, isolated from the bark of Pacific yew trees, used in the treatment of breast cancer.

Medical researchers hoped for a long time that, instead of relying on trial and error, new drugs could be designed on the basis of a rational understanding of the body's physiology and molecular biology. The discovery of vitamins and the identification of hormones like insulin were important steps in this direction, and from the 1980s onwards, there were great hopes that the understanding of genomes and the molecular details of cells would take 'rational' drug discovery to a new level. With this aim in view, hundreds of billions of dollars have been invested by governments, drug companies and biotechnology firms. But the results have been very disappointing. The returns on investment are diminishing, and drug companies are now facing a dearth of new drugs. At the same time, the patents on some of the main 'blockbuster' drugs like Lipitor, a statin for the control of cholesterol levels, and Prozac, an antidepressant, are running out, meaning a loss of many billions of dollars in annual revenues for pharmaceutical corporations. Many of the new drugs now in the pipeline are merely more expensive variants of already-existing drugs.[10]

Discovering and testing new drugs is a lengthy and increasingly expensive process, and drug companies try to make as much money as possible from their drugs while patents last. They inevitably devote enormous amounts of money to advertising and promotion. Some companies go to great lengths to make their drugs look safer and more effective than they really are, creating an illusion of scientific respectability for their claims. In order to bolster the drugs' scientific credibility, they offer large fees to scientists to put their names to articles that have been ghostwritten by authors paid by the drug company, or else the scientists are given other inducements to lend their names to studies they have not done.[11]

Medical ghostwriting takes many forms, but a recent case gives some insight into what is involved. In 2009, around fourteen thousand women who developed breast cancer while taking Prempro, a hormone replacement therapy (HRT), sued the drug's manufacturer, Wyeth. In court, it turned out that many of the medical research papers supporting HRT had been ghost-written by a commercial medical communications company called DesignWrite, whose website boasted that over twelve years they had 'planned, created and/or managed hundreds of advisory boards, a thousand abstracts and posters, 500 clinical papers, over 10,000 speakers' bureau programmes, over 200 satellite symposia, 60 international programmes, dozens of websites, and a broad array of ancillary printed and electronic materials'.[12] It emerged that DesignWrite organised a 'planned publication programme' for Prempro, consisting of review articles, case reports, editorials and commentaries, using the medical literature as a marketing tool. As Ben Goldacre reported in the *Guardian*,

> DesignWrite wrote the first drafts and sent them to Wyeth, who advised on the creation of a second draft. Only then was the paper sent to the academic who would appear as the 'author' . . . DesignWrite sold Wyeth more than 50 peer-reviewed journal articles for HRT, and a similar number of conference posters, slide kits, symposia, and journal supplements. Adrienne Fugh-Berman [an associate professor of physiology at Georgetown University] found that these publications variously promoted unproven and unlicensed benefits of Wyeth's HRT drug, undermined its competitors, and downplayed its harms . . . [A]cademic journal publications are not regarded as promo-tional activity, so all this was legal. Worst of all was the complicity of the academics . . . 'Research shows high clinician reliance on journal articles for credible product information,' said DesignWrite. They're right: when you read an academic paper, you trust it was written by the person whose name is on it.[13]

Pharmaceutical companies also have a strong influence on govern-ments and on the public funding of medical research. In the United States, between 1998 and 2004, drug companies and their trade groups, the Pharmaceutical Research and Manufacturers of America (PhRMA) and the Biotechnology Industry Organization, spent more than $900 million on lobbying, including donations of $90 million to political parties and election campaigns, mainly Republicans. They lobbied on at least 1,600 pieces of legislation, with more than 1,200 registered lobbyists in Washington, DC.[14]

In the UK, the official Medicines and Healthcare Products Agency, which regulates the pharmaceutical industry, is funded by the pharmaceutical industry itself. The funders inevitably influence the regulator's actions. For example, in February 2008, the Agency decided that in the light of recent evidence, a new warning about side effects should be put on the label of statins. But nothing happened for twenty-one months because one of the drug companies 'was not in agreement with the wording'. As Ben Goldacre commented in the *Guardian*, 'A drug company has been able to delay the inclusion of safety warnings on a drug prescribed to 4 million people for 21 months because it didn't agree with the wording. There is no conceivable world in which this is a good thing.'[15]

Sometimes drug companies simply ignore the regulatory process and sell drugs 'off-label', in other words for uses that have not been approved because the drug has not been shown to be safe, necessary and effective. A flagrant case came to light in 2010, when the US Justice Department fined AstraZeneca $520 million for the 'off-label' marketing of its blockbuster anti-psychotic Seroquel. This drug was approved only for the short-term treatment of schizophrenia and acute bipolar disorder, but for five years AstraZeneca had been aggressively marketing the drug as a long-term cure-all, promoting its use in old people's homes, veterans' hospitals and prisons, as well as for the treatment of agitation and aggression in children, even though clinical studies had shown 'serious and debilitating side effects', particularly among the elderly and in children.[16] The same company was fined $355 million in 2003 for the fraudulent selling of Zoladex, a prostate cancer treatment. Although these fines were among the largest imposed by the US Justice Department on pharmaceutical companies, critics point out that the fines were less than 20 per cent of the revenues from off-label marketing. By settling out of court the companies avoided criminal convictions, no one went to jail, and the fines were treated as a cost of doing business.[17]

Obviously it is in the interests of pharmaceutical companies to sell as many expensive drugs as possible, even though the interests of patients and those who pay for their healthcare are different. This conflict of interest needs to be mediated by governments, independent regulatory agencies, and independent researchers. Unfortunately, the lobbying of governments, financial control of regulatory agencies and the funding of medical researchers by the industry means that pharmaceutical corporations have a huge influence on the entire medical system, and reinforce its reliance on drugs.

Placebo responses and the power of hope

How much of the success of more drugs is really due to the drugs themselves, and how much depends on people's beliefs and expectations?

In scientific and medical research, as in everyday life, our beliefs, desires and expectations can influence, often subconsciously, how we observe and interpret things.[18] There is overwhelming experimental evidence that scientists' attitudes and expectations can influence the outcome of experiments.[19] In experimental psychology and clinical research, these principles are widely recognised, which is why experiments in these subjects are often carried out under 'blind' conditions.

In medicine, patients' expectations also influence the results, and double-blind procedures are used to guard against the expectations of both subjects and investigators. For example, in a typical double-blind clinical trial of a drug, some patients, selected at random, are given tablets of the drug being tested and others are given similar-looking placebo tablets, pharmacologically inert. The purpose of these trials is to find out if the new drug works better than the placebo. Only if it does so can it be licensed and marketed as an effective treatment. Neither clinicians nor patients know who gets which. In such experiments, the placebo often works in a similar way to the drug being tested, although usually to a lesser extent.

The largest placebo responses tend to occur in trials in which both patients and physicians believe a powerful new treatment is being tested. The blank pills work because the patients who take them and the doctors who administer them think that they might contain the new wonder drug.[20] If the trial is not blind, and the patients and doctors know who has been given the real drug and who has the placebo, the placebo response is greatly reduced. Neither the patients nor the doctors expect the placebo to have much effect, and it doesn't.[21] This can be a serious problem even in double-blind trials. If the real drug has noticeable side effects, both the patients and the doctors can find out who is receiving the placebo and the real drug, and as a result the placebo is less effective, which makes the real drug more effective relative to the placebo.[22] This may seem like a tiresome technical detail, but it has huge economic consequences.

For example, in several clinical trials, the antidepressant drug Prozac had slightly more effect than the placebo, and was licensed for use, resulting in annual revenues to the manufacturers of more than $2 billion. But was it really better than a placebo? Perhaps not. Although the trials were double-blind, Prozac has some well-known side effects, such as nausea and insomnia. Both patients and clinicians might have realised who had received the Prozac and who the placebo by noticing these side effects, or their absence. This is called 'breaking the blind'. Once some people realised they were receiving the real drug and others realised they were getting the placebo, the placebo would have become less effective, and hence Prozac would have seemed more effective by comparison. In a study in which doctors and patients were asked to say whether they had received the real drug or the placebo, 80 per cent of the patients and 87 per cent of the doctors were right, as opposed to the 50 per cent that would be expected by random guessing.[23]

However, in several other clinical trials, Prozac was no better than the placebo. One reason could be that in these trials the patients had less experience with antidepressants and were less able to recognise the side effects. However, the drug company, Eli Lilly, did not publish the results of unsuccessful trials, which were revealed only because an independently minded researcher, Irving Kirsch, managed to obtain the data using the US Freedom of Information Act. He found that when all the data were taken into account, not just the positive results published by the manufacturers, Prozac and several other antidepressants turned out to be no more effective than placebos, or than a herbal remedy, St John's wort, which is far cheaper.[24] Ironically, the suppression of the data showing that Prozac was no better than a placebo probably helped to increase its effectiveness as a prescription drug, because doctors and patients had more belief in it, thus enhancing the placebo response.

Blind assessment first began in the late eighteenth century as a tool for detecting fraud. Mainstream scientists and physicians invented blind methods in order to challenge the suspected charlatanism of unconventional medicine.[25] Some of the first experiments of this kind were carried out to evaluate mesmerism, and were literally conducted with blindfolds. They took place in France at the house of Benjamin Franklin, the American representative in Paris, who was head of a commission of enquiry appointed by King Louis XVI. Homeopaths adopted blind assessment in the mid-nineteenth century, and psychologists and psychical researchers took up blind methodologies before 1900 to prevent the subject's beliefs and expectations influencing their responses. But in regular medicine, blind methods were rarely used until the 1930s. It was only after the Second World War that double-blind trials to compare drugs and placebos became a standard technique for medical researchers.

Although the word 'placebo' usually conjures up the image of an inert sugar pill, any treatment that patients believe will make them better can evoke a placebo response, even fake surgery. In the 1950s, many surgeons carried out an operation to relieve angina, a severe chest pain caused by lack of blood in the heart muscle. They tied off some of the arteries that carried blood to the chest. In a placebo-controlled study, some patients received only sham surgery, in which their chests were cut open and stitched up again. Much to their surprise, the doctors found that the dummy operation did almost as much good as the real one. The mere belief that they had received a proper operation relieved the patients' chest pain.[26]

In a similar way, injecting people with saline solution often brings about cures, even though no drug is used. Placebo injections are especially effective when people have a strong belief that injections are magically powerful, as in rural Africa and Latin America.[27] Placebo injections also give bigger placebo responses than placebo pills in the United States, but in Europe they do not.

Placebo responses depend on the meanings that people attribute to diseases and cures,[28] and these vary from culture to culture, as the research of medical anthropologists has shown. For example, in a comparison of clinical trials in a wide range of countries, Germans had the highest placebo healing rate for ulcers and the lowest for high blood pressure.[29] One possible reason for the lower placebo response of Germans in blood-pressure trials is that Germans are unusually concerned with the heart and its workings. Although the rate of heart disease is the same in Germany, France and England, Germans take about six times as much heart medication as their neighbours, and German physicians are almost alone in prescribing medication for *low* blood pressure. Worries by German patients about their blood pressure getting too low may have reduced the placebo response in trials of drugs to lower blood pressure compared with patients in other countries without this concern.[30]

For many years, most medical researchers regarded placebo responses as an annoying complication in clinical trials. They got in the way of finding out about real cures. But attitudes are changing. The placebo response shows that patients' beliefs and hopes play an important part in the healing process.

At first, the defenders of mechanistic medicine dismissed the effects of complementary and alternative therapies as 'mere' placebo effects. But placebo responses play an important part in conventional medicine too. As Simon Singh and Edzard Ernst observed,

[T]he impact of a proven treatment is always enhanced by the placebo effect. Not only will the treatment deliver a standard benefit, but it should also deliver an added benefit because the patient has an expectation that the treatment will be effective . . . The best doctors fully exploit the placebo impact, while the worst ones add only a minimal placebo enhancement to their treatments.[31]

In 2009, placebo responses turned out to be increasing – especially in the United States. In clinical trials, fewer and fewer new drugs beat the placebos. In other words, more and more drugs failed in clinical trials, causing big problems for drug companies.

Why have placebo responses increased in the United States, but not elsewhere? The answer may be that drug companies are victims of their own success. In 1997, direct-to-consumer advertising for drugs was made legal in the United States, and as a result US citizens have been deluged with advertisements for prescription medicines. Many of these commercials evoke uplifting associations between pills and peace of mind. The pharmaceutical industry's advertising has been all too successful in raising expectations about new drugs, increasing the placebo response in clinical trials, and hence reducing the difference between the placebo and the drug being tested.[32]

If materialism were an adequate foundation for medicine, placebo responses ought not to occur. The fact that they do occur shows that people's beliefs and hopes can have positive effects on their health and healing. Conversely, despair and hopelessness can have negative effects. There is even a field of research devoted to this subject: psychoneuroimmunology. Stress, anxiety and depression suppress the activity of the immune system, and make it less able to resist diseases and inhibit the growth of cancerous cells.[33] Hence people who are anxious or depressed are more likely to fall ill or get cancer.

Placebo responses show that health and sickness are not just a matter of physics and chemistry. They also depend on hopes, meanings and beliefs. Placebo responses are an integral part of healing.

Hypnotic blisters and the removal of warts

Through suggestion, one person can guide the thoughts or feelings of another. This is a normal part of everyday life. But 'the power of suggestion' can bring about exceptionally striking effects through hypnosis. The nature of hypnosis has been debated for decades, but there is no doubt that it occurs, and produces visual illusions and other subjective effects. But hypnosis can also affect the body.

When I was studying at Cambridge, one of our physiology lecturers, Fergus Campbell, gave a demonstration of the powers of hypnosis using one of my fellow students as a subject. Campbell told the subject that he was carrying out a scientific experiment on the response of skin to heat, and would be touching the subject's arm with a lighted cigarette. In fact he touched it with the flat end of a pencil. Soon afterwards, the skin reddened and a blister appeared where the cool pencil had touched. I later learned that many other hypnotists had shown the same thing, and that it had been studied, but not explained, by medical researchers.[34]

Nerves that control small arteries in the skin mediate this burn response. People cannot will themselves to activate these nerves, which are under the control of the autonomic or involuntary nervous system. Yet the hypnotic induction of burns shows that suggestion can work through the autonomic nervous system. Functions that are normally involuntary are potentially subject to mental influence.[35] This same principle is also demonstrated by biofeedback training. For example, in one common practice, people learn to increase the blood flow to their hands by paying attention to the temperature in their fingers, which is converted to an audio or visual display, so that they receive continuous feedback. If the temperature is indicated by the rate at which they hear clicks, their task is to speed up the clicks. Without knowing how, most people can soon learn to increase the blood flow to their fingers, and hence raise their temperature. With practice they can do this on their own without the machine.[36]

Hypnosis can also produce 'miracle cures', as it did in the case of a boy in London in the 1950s. He was born with a thick, dark skin, and as he grew, most of his body was covered with a black, rough casing. Doctors said he had been born with ichthyosis, 'fish-scale disease'. Treatment at some of London's best hospitals did no good. Even a skin transplant from his normal chest to his hands proved worse than useless: the skin blackened and then shrank, stiffening his fingers. Albert Mason, a young doctor interested in hypnosis, heard of the case and, watched by a dozen sceptical colleagues, put the boy into a hypnotic trance. He told him, 'Your left arm will clear.' It did. About five days later the coarse outer layer of skin became crumbly and fell off. The skin underneath soon became pink and soft. Through repeated hypnosis, Mason cleared other parts of the body, limb by limb.[37] In a follow-up study three years later, Mason and a team of dermatologists confirmed that 'not only has there been no relapse, but his skin has continued to improve'.[38]

Mental influences are often effective in wart cures. Warts on the skin are made up of abnormal tissues infected with viruses. Conventional doctors usually treat them by cutting them off with knives, or burning them with electric sparks, or freezing them with liquid nitrogen, or dissolving them with corrosive acid. These methods are crude, sometimes painful, and often ineffective: in many cases the warts grow back, sometimes in multiple clusters. Yet 'miracle' cures can work much quicker and more effectively. Some people build up reputations as 'wart healers', and cure warts just by touching them. Others do it by applying curative plants. Another method is to rub the wart with a potato and then bury the potato under a particular tree at a particular phase of the moon. Some people get rid of their warts by selling them to a sibling. Often, within a few days of one of these treatments, the wart drops off, leaving clear skin. Or sometimes it gradually shrinks and disappears over a week or two.[39]

The 'magical' methods for curing warts are many and various. They cannot have significant direct effects on the viruses or the abnormal tissue, but they provide rapid, lasting healing. All that they have in common is belief. The person with the wart hopes that the method will work, and it often does.[40]

The effects of lifestyle, social networks and spiritual practices

Everyone agrees that health is affected by people's habits and life-styles. The causative role of smoking on lung cancer provides the clearest example. Until the 1950s most people did not recognise that smoking had harmful effects, and the epidemiological research that established the facts is one of the major achievements of modern medicine. For example, a large-scale study of British doctors began in 1953 by documenting their smoking habits and over the subsequent decades recording their mortality. This was an example of what researchers call a prospective, as opposed to retrospective, study in which groups identified at the beginning of the study are tracked over time. It turned out that those who smoked more than twenty-five cigarettes a day had a twenty-five-fold greater risk of dying from lung cancer than non-smokers.[41]

Anti-smoking education, restrictions on advertising cigarettes and bans on smoking in public places have contributed to a fall in the percentage of people who smoke and to a decline in the incidence of lung cancer. In men in the UK, the incidence of lung cancer peaked in the late 1970s, and by 2011 had fallen by more than 45 per cent. Emboldened by this success, from the 1980s onwards, health policy-makers embraced the 'social theory' of disease, initially in relation to heart disease, and more recently in connection with the epidemic of obesity and its associated health disorders. They have rightly emphasised the importance of a healthy diet and exercise, and some people have changed their lifestyles accordingly. Yet many have not.[42] Obviously many factors influence these trends, including sedentary lifestyles, junk foods and sugary drinks, and increases in obesity are now occurring in many other parts of the world. More than 1 billion people are now estimated to be overweight, including more than 300 million who are clinically obese. Exhortations by the medical profession and by governments have failed to reverse this trend.

The social and economic aspects of medicine show that the materialist model of people as machines is much too limited.

People's motivations and attitudes, the effects of social networks and the influence of advertising are not measurable physical and chemical forces: they work through minds. Many other lines of evidence show that health is influenced by social, spiritual and emotional factors. For example, in studies in the United States, men who had suffered a heart attack were four times more likely to die in the next three years if they were socially isolated. Both men and women who had undergone coronary operations were three times more likely to survive for five years if they were married or had a close friend.[43] Other studies showed that more people who kept pets survived after heart attacks than those without, and elderly and bereaved people who kept dogs or cats had better health and needed less medication than those who did not have animals to keep them company.[44]

Numerous studies in the United States and elsewhere have shown that people who are religious, especially those who regularly attend religious services, live significantly longer, and have better health and less depression than people without religious faith. These effects were found with Christian and non-Christian groups.[45] Some of the benefits may be a result of community support and other social factors, but the spiritual practices themselves may also be important.

The effects of prayer or meditation on health and survival have been investigated through prospective studies in which people who prayed or meditated and otherwise similar people who did not pray or meditate were identified at the start of the study and watched over a period of years to see if their health or mortality turned out differently. It did. On average, those who prayed or meditated remained healthier and survived longer than those who did not.[46] For example, in a study in North Carolina, Harold Koenig and his colleagues tracked 1,793 subjects who were over sixty-five years old with no physical impairments at the beginning of the study. Six years later, those who prayed had survived 66 per cent more than those who did not pray, after correcting for age differences between the two groups. (Without this correction the difference was 73 per cent.) They then examined the effects of 'confounding variables', a scientific term for other factors that might have influenced survival, like stressful life events, depression, social connections and healthy lifestyles. Even after controlling for these variables, those who prayed survived 55 per cent more. 'Thus, healthy subjects who prayed were nearly two-thirds more likely to survive, and only a small percentage of this effect could be explained on the basis of mental, social, or behavioral factors.'[47]

If a new drug or surgical procedure had such dramatic effects on health and survival as spiritual practices, it would be hailed as a medical breakthrough.

Changes in official thinking

In an article in *Nature* in 2011, Michael Crow, the president of Arizona State University and a senior science administrator, proposed a radical overhaul of the US National Institutes of Health, most of whose $30 billion annual budget is directed towards the discovery of molecular and genetic aspects of diseases rather than looking at people's behaviour. He proposed that the present 'Byzantine array' of twenty-seven institutes and centres be replaced with three new institutes. One would look at fundamental questions relating to human health, including sociological and behavioural perspectives. A second would be devoted to research on health outcomes, defined by measurable improvements in people's health:

> This should draw on behavioural sciences, economics, technology, communications and education as well as on fundamental biomedical research . . . If the aim is to reduce national obesity levels – currently around 30% of the US population is obese – to less than 10% or 15% of the population, for example, project leaders would measure progress against that goal rather than according to some scientific milestone such as the discovery of a genetic or microbial driver of obesity.[48]

The third of the new institutes would be for health transformation: 'Rather than being rewarded for maximising knowledge production, this institute would receive funding based on its success at producing cost-effective public health improvements.'[49]

No doubt attempts to change people's behaviour will be controversial politically and will come into conflict with powerful financial interests, such as those of the food and drink industries. But public-health problems seem unlikely to be soluble by drugs or surgery alone, and medical costs related to obesity, estimated to be about $160 billion a year in the United States in 2011, are projected to double by 2020.[50]

Similar changes in thinking are occurring in other countries. In 2010, the UK government published an official report on health policy, a White Paper entitled *Healthy Lives, Healthy People*, which strongly emphasised social factors that affect health and disease. As in the United States, the economic issues were in the foreground, especially in relation to seemingly voluntary aspects of health and disease. The Minister for Health, Andrew Lansley, wrote in the Foreword:

> We have to be bold because so many of the lifestyle-driven health problems we see today are already at alarming levels. Britain is now the most obese nation in Europe. We have among the worst rates of sexually transmitted infections recorded, a relatively large population of problem drug users and rising levels of harm from alcohol. Smoking alone claims over 80,000 lives every year. Experts estimate that tackling poor mental health could reduce our overall disease burden by nearly a quarter . . . We need a new approach that empowers individuals to make healthy choices.[51]

The fact that influential administrators and government ministers are proposing radical reforms is a sign of a general change in attitude to health and disease, a swing away from a focus on drugs and surgery to a social model that takes into account people's behaviour and motivations, as well as motivational and economic factors that lie outside the scope of old-style mechanistic medicine.

Complementary and alternative therapies

One of the paradoxes of modern medicine is that, despite its great triumphs and successes, from the 1980s onwards there has been an enormous surge in the popularity of alternative therapies, previously of interest to only a small minority and widely perceived as fraudulent. Part of the reason may be that many alternative practitioners spend more time with their patients and take a more personal interest in them than orthodox doctors, who are working under a greater pressure of time. Another reason may be that doctors' preoccupation with drugs has led to the neglect of simpler, more traditional remedies and the dismissal of anything that does not fit with the mechanistic conception of disease. For example, as James Le Fanu has pointed out, in relation to problems concerning the joints, muscles and bones,

> following the discovery of cortisone and other anti-inflammatory agents, the skills of rheumatologists devolved around juggling various toxic regimes of drugs in the hope that the benefits might outweigh the sometimes grievous side-effects. Meanwhile, all the other therapies for rheumatological disorders – such as massage, manipulation and dietary advice – were abandoned virtually wholesale, only to be 'rediscovered' by alternative practitioners in the 1980s.[52]

There are many different alternative and complementary therapies. Some, like homeopathy, naturopathy and chiropractic, grew up in the nineteenth century in opposition to the practice of orthodox medicine, which was often harmful; standard procedures included bleeding patients through incisions or with leeches. Then there were various mind or faith cures, including miraculous healing at Catholic shrines like Lourdes, faith healings by Protestant evangelists, and Christian Science, a church founded in the United States by Mary Baker Eddy (1821–1910), who taught that disease, injury, pain and even death were illusions that were given their power by minds out of harmony with God. In response, official doctors often opposed these rival systems and denounced them as dangerous quackery.[53] In addition to the wide range of home-grown alternative systems in the West, there are now many therapists who practise traditional medical systems from other parts of the world, including shamanic healing rituals, ayurvedic medicine from India and traditional Chinese medicine, including acupuncture.

Most of these alternative practices are based on non-materialist systems of thought, and therefore dogmatic materialists see them as superstitious or fraudulent. Yet all these systems claim to have cured people. Some have had impressive levels of success in clinical trials, suggesting that they 'really' work. For example, in 2003, the World Health Organization published a review of 293 controlled clinical trials of acupuncture, and came to the conclusion that acupuncture was an effective treatment for a wide range of conditions, including morning sickness and stroke.[54] This evidence was inevitably controversial. For people who believed that real effects are impossible, then the evidence *must* have been flawed. For example, critics argued that all the acupuncture trials conducted in China should be excluded because the results were too positive.[55] Nevertheless, critical reviews of non-Chinese studies also showed positive effects of acupuncture; for example, in the relief of pain and nausea.[56] But disputes rumble on, because in acupuncture trials it is impossible to conduct truly double-blind trials. The acupuncturist has to know whether he or she is giving 'placebo' acupuncture using fake needles.

Nevertheless, everyone is prepared to admit that alternative therapies can work as placebos. And since placebo responses themselves really work, this raises the question as to whether some therapies work better than others, even if they are indeed placebos. Some may bring about bigger placebo responses and hence heal people more effectively.

Evidence-based medicine and comparative effectiveness research

It is often assumed that the only scientifically valid kind of clinical trial is a randomised double-blind placebo controlled study, the 'gold standard' methodology. These trials are indeed helpful in distinguishing between the effects of a treatment and the effects of a placebo, but they do not provide the information that is needed by many patients and health-care organisations. For example, if I am suffering from lower back pain, I do not want to know whether drug X works better than a placebo in relieving this condition, but which kind of treatment I should seek among the various available therapies, mainstream and alternative: physiotherapy, or drugs from my doctor, or acupuncture, or osteopathy, or some other therapy.

The best way to answer this question is by comparing the outcomes of different kinds of treatment, doing the trial as fairly as possible on a level playing field. The question would be purely pragmatic: what works? For example, equal numbers of sufferers with lower back pain could be allocated at random to a range of treatment methods, including physiotherapy, osteopathy, chiropractic, acupuncture, and any other therapeutic methods that claim to be able to treat this condition; there would also be a group that is given no treatment at all, by being put on a waiting list. Within each treatment group, there would be several different practitioners, so that not only could the different methods be compared, but also the variability between practitioners of any particular method.

The outcomes would be assessed in the same way for all patients at regular intervals after the treatment. The relevant outcome measures would be agreed in advance in consultation with the therapists involved. The data would then be analysed statistically to find out

1. Which treatment, if any, worked best.
2. Which treatment methods had the greatest variability between practitioners.
3. Which methods were the most cost-effective.

This kind of information would be of great use to patients and also to providers of health care, such as the UK National Health Service. A similar level-playing-field approach could be adopted for a variety of other common conditions, including migraine headaches and cold sores. This kind of research, sometimes called Comparative Effectiveness Research, would be relatively simple and cheap to conduct.

Imagine, for example, that homeopathy turned out to be the best treatment for cold sores. Sceptics would argue that this was simply because homeopathy had a stronger placebo effect than the other treatments. But if homeopathy did indeed unleash a greater placebo response, then this would be an advantage, not a disadvantage. Homeopathy would really work, and it would probably be cheaper too.

Outcome research of this kind is already used within medicine to a limited extent, especially in relation to mental disorders, like depression and schizophrenia. Although many psychiatrists and the pharmaceutical industry regard modern antidepressant and anti-psychotic drugs as 'curing' chemical imbalances in the brain, others argue that, instead, these drugs work because they are psychoactive drugs, rather than specific cures; they alter states of mind, with effects that include the suppression of emotions and intellectual activity.[57] The drugs are useful, but they are not chemical cures. By contrast, psychotherapy has more lasting effects, whether it is combined with drugs or not. There have already been hundreds of outcome studies on the treatment of depression by psychotherapy rather than drugs, and the results are clear. Irving Kirsch summarised them:

> Psychotherapy works for the treatment of depression, and the benefits are substantial. In head-to-head comparisons, in which the short-term effects of psychotherapy and antidepressants are pitted against each other, psychotherapy works as well as medication. This is true regardless of how depressed the person is to begin with . . . Psychotherapy looks even better when its long-term effectiveness is assessed. Formerly depressed patients are

far more likely to relapse and become depressed again after treatment with antidepressants than they are after psychother-apy.[58]

These important conclusions would not be possible unless the effectiveness of different kinds of treatments had been compared. It would never emerge from research that concentrated only on placebo-controlled trials of drugs.

One of the problems with mechanistic medicine is its tunnel vision and its obsession with chemical and surgical methods to the exclusion of all others. For decades, the materialist worldview has shaped the way that medicine is taught in medical schools, biased the funding of medical research, and shaped the policies of national health services and private insurance companies. Meanwhile medicine has become ever more expensive.

Comparative effectiveness research could lead to a genuinely evidence-based system of medicine that would include, rather than exclude, therapies that do not fit with the materialist belief-system.

Fantasies of immortality

Most people, like most doctors, are pragmatic, but there is tension at the heart of modern medicine between realistic expectations of what science and medicine can do and the dream of physical immortality. For those who have turned scientific progress into a kind of religion, the scientific conquest of death becomes the ulti-mate goal. Alchemists failed to discover the legendary elixir that was supposed to confer eternal life or eternal youth, but some of the wilder enthusiasts for the scientific salvation of humanity believe that science itself will enable some humans to live for ever.

The idea of physical immortality was surprisingly influential in the early years of the Soviet Union, when some of the more vision-ary intellectuals were obsessed with the idea of 'God-building'. Through science, humanity would become all-powerful. Man would become godlike and abolish physical death.[59] One of these so-called God-builders was Leonid Krasin (1870–1926), the People's Commissar of Foreign Trade in Lenin's government. In 1921, three years before Lenin's death, he asserted, 'That time will come, when the liberation of mankind, using all the might of science and technology . . . will be able to resurrect great historical figures.'[60]

When Lenin died, he was first embalmed and then frozen, using a system Krasin designed. An official commission, called 'The Immortalisation Commission', oversaw the building of Lenin's mausoleum, which became a place of pilgrimage for Communists, just as the shrines of saints had been places of pilgrimage for Christians. But Lenin's corpse decayed, despite all Krasin's efforts.

Today, in the United States, several companies offer more advanced refrigeration systems for the same purpose. In 2011, the price for whole-body preservation in liquid nitrogen was around $150,000. 'Neuropreservation' was cheaper: severed heads were frozen for about $90,000.[61] Six companies offer this service, and dozens of Americans are already frozen and awaiting resurrection.

Freezing is merely a stopgap measure, and some people hope that death itself will soon be overcome. In 2009, the futurist Ray Kurzweil claimed that humans could be immortal in as little as twenty years' time, thanks to nanotechnologies and nano-robots, or 'nanobots', that would enable vital organs to be replaced:

> I and many other scientists now believe that in around 20 years we will have the means to reprogram our bodies' stone-age software so we can halt, then reverse, ageing. Then nanotechnology will let us live for ever. Ultimately, nanobots will replace blood cells and do their work thousands of times more effectively. Within 25 years we will be able to do an Olympic sprint for 15 minutes without taking a breath, or go scuba-diving for four hours without oxygen . . . If we want to go into virtual-reality mode, nanobots will shut down brain signals and take us wherever we want to go. Virtual sex will become commonplace.[62]

Meanwhile, to delay the ageing process so that he can survive long enough to benefit from these advances, Kurzweil takes 250 supplement pills a day.[63] But unless and until his dreams come true, we will all have to die of something, and the longer death is delayed, the more expensive and medically demanding our lives will become.

Most doctors take a pragmatic view of their abilities, and recognise that there are limits to the power of medicine. A conquest of one disease, or at least its diminution, must inevitably increase the death rate from other diseases. If all heart disease could be prevented or cured, then death rates from cancer would go up. If all cancers could be cured, then death rates from other causes would increase. And as new drugs and new techniques become ever more expensive, and as more people survive to old age, the costs of treatment are becoming increasingly unaffordable, even in the richest countries.

Ways of dying

Surgeons can operate on people with lung cancer, but they cannot stop people smoking and increasing their likelihood of getting lung cancer in the first place. They can operate on old people to replace failing organs, but such operations become increasingly risky and expensive, and give a very limited extension of life. In the United States, about 30 per cent of the Medicare budget, which pays the health expenses of people over sixty-five, is spent on patients in their last year of life; and 78 per cent of this expenditure is in the last month of life.[64]

A study funded by the US National Cancer Institute compared alternative ways of treating patients with advanced cancer. Some were treated in the standard way, without being asked about their preferences. Others had an 'end of life' conversation with their physician, in which one of the questions was: 'If you could choose, would you prefer (1) a course of treatment that focused on extending life as much as possible even if that meant more pain and discomfort, or (2) a plan of care that focused on relieving pain and

discomfort as long as possible even if that meant not living as long?'
Many of the patients preferred the second option: they did not
want to die on a respirator in intensive care. Patients who were
given this choice 'had significantly lower health care costs in their
final week of life. Higher costs were associated with worse quality
of death.'[65] In another study, patients with metastatic lung cancer
who were given palliative care soon after diagnosis reported better
quality of life, were less depressed and, on average, actually survived
longer than those who received aggressive anti-cancer therapy.[66]

Hospices and palliative care provide a very different way of
approaching death. Palliative care focuses on relieving and
preventing suffering. Instead of seeing terminal illness as a medical
crisis requiring extreme interventions, patients are cared for in a
way that helps them prepare for death, emotionally, socially and
spiritually, as well as physically.

What difference does it make?

At present we have an official state-sponsored medical system that
is expensive, restrictive and strongly influenced by powerful pharmaceutical corporations, whose primary concern is the making of
large profits. This system has been spectacularly successful, but
most of its advances occurred before the 1980s. The rate of innovation is slowing, and most of the promises of genetic medicine and
biotechnology remain unfulfilled. Meanwhile the costs of treatment and of research are increasing.

If the state-sponsored monopoly of materialism is loosened,
scientific and clinical research could look at the role of beliefs,
faiths, hopes, fears and social influences in health and healing.
Systems of therapy could be compared on the basis of their effectiveness, and people could choose those that are likely to work
best for them, with the help of informed advisers. Diet, exercise
and preventive medicine programmes would also be compared on
the basis of their effectiveness. The nature of placebo responses
and the power of the mind could become valid fields of research,
as would the effects of prayer and meditation.

An integrative medical system could empower people to lead healthier lives. Doctors and patients could become more aware of the innate capacity of the body to heal and could recognise the importance of hope and faith. More people could be asked how they would prefer to die, whether at home, in a hospice or in intensive care.

An integrative approach to medicine would build on the enormous advances of the last two centuries, and include them in a broader kind of medicine that could work better and cost less.

Questions for materialists

Have you ever consulted an alternative therapist? If not, would you ever consider doing so?

How do you explain the placebo response?

How do you think governments and insurance companies should deal with the escalating costs of medicine?

Do you think governments should fund comparative effectiveness research on different kinds of therapy, including alternative therapies?

SUMMARY

Modern medicine has been amazingly successful. Together with immunisation programmes and public-health measures it has reduced infant mortality, transformed human lives and increased life expectancy. Its focus on the physical and chemical aspects of human bodies has resulted in major advances in surgery and drugs. But because of its materialist prejudices, it ignores mental influences as much as possible. People's hopes and expectations affect their recovery from disease, injury or surgery, as revealed quantitatively in placebo responses. The power of belief is also shown by the hypnotic induction of blisters and by 'magical' cures for warts. Conversely, feelings of despair and hopelessness can suppress the activity of the immune system, lead to poorer rates of recovery from injuries and surgery. People who have suffered from heart attacks survive better, on average, if they are married, or have a close friend, or keep a pet. Regular attendance at religious services tends to lead to better health and longevity, and people who pray or meditate tend to be healthier and live longer than those who do not. Thus many psychological, emotional, social and spiritual factors affect health and disease. So do diet and lifestyle. The 'obesity pandemic' and the spiralling costs of healthcare are forcing changes in government policies, but exhortation and education are of limited effectiveness in changing people's motivation and behaviour. Alternative and complementary systems of medicine cure some people some of the time, and not all their effects can be ascribed to placebo responses alone. Comparative effectiveness research provides a way of finding out what works best. All medical systems involve placebo responses, and some may produce more than others. When people are nearing death, heroic attempts to keep them alive by emergency surgical interventions are expensive and often inappropriate. If they are given the choice, many prefer palliative care and prefer a hospice to a hospital, even if they are likely to die sooner. An inclusive, integrative medical system is likely to be cheaper and more effective than an exclusively mechanistic system.

11

Illusions of Objectivity

For those who idealise science, scientists are the epitome of objectivity, rising above the sectarian divisions and illusions that afflict the rest of humanity. Scientific minds are freed from the normal limitations of bodies, emotions and social obligations, and can travel beyond the earth-bound realm of the senses to see all nature as if from outside, stripped of subjective qualities. They have godlike, mathematical knowledge of the vast reaches of space and time, and even of countless universes beyond our own. Unlike religion, locked in endless conflicts and disputes, science offers a true understanding of material nature, the only reality there is. Scientists constitute a priesthood superior to the priesthoods of religions, which maintain their prestige and power by playing on human ignorance and fear. Scientists stand in the vanguard of human progress, leading humanity onwards and upwards to a better and brighter world.

Most scientists are unconscious of the myths, allegories and assumptions that shape their social roles and political power. These beliefs are implicit rather than explicit. But they are more powerful because they are so habitual. If they are unconscious, they cannot be questioned; and in so far as they are collective, shared by the scientific community, there is no incentive to question them.

In the course of this book, I have shown that the materialist philosophy or 'the scientific worldview' is not a vision of undeniable, objective truth. It is a questionable belief-system superseded by the development of the sciences themselves. In this chapter I look at the myths of disembodied knowledge and scientific objectivity and the ways in which they conflict with the obvious fact that scientists are people. Sciences are human activities. The assumption that the sciences are uniquely objective not only distorts the public perception of scientists, but affects scientists' perception of themselves. The illusion of objectivity makes scientists prone to deception and self-deception. It works against the noble ideal of seeking truth.

Shamanic journeys and disembodied minds

From the outset, the power of science to persuade has depended not only on quantitative calculations, reason and power, but on the use of the imagination. This is illustrated nowhere more clearly than in a remarkable book by Johannes Kepler, written in 1609, called *Somnium, sive astronomia lunaris*, meaning 'The Dream, or the Astronomy of the Moon'. The aim of his book, he explained, was to 'work out, through the example of the moon, an argument for the motion of the earth'.[1] One of the biggest problems faced by Kepler and other defenders of the idea that the earth moved around the sun – in other words the Copernican system of astronomy – was that the earth feels stationary and we actually see the sun moving around it.

In his *Somnium*, Kepler depicted a journey to the moon and described the universe as seen from its surface. The moon 'seems to its inhabitants to be stationary, while the stars go around it, just as the earth seems to us to be stationary'.[2] His voyager saw the earth hanging in space and revolving on its axis. Thus, by imagining a journey to the moon, he made the new astronomy imaginable. This vision was embodied in globes. Everyone who sees a globe in a geography classroom can experience the earth seen from a point of view outside it, which no human being had ever actually seen until the first astronauts looked back at the earth from space. But the extraterrestrial point of view long preceded the Copernican revolution. Greek astronomers had already concluded by the third century BC that the earth was spherical, and were making globes.[3] The novel feature of Kepler's vision was not seeing the earth from outside, but seeing it rotating.

Kepler's observer was able to go to the moon because he was a disembodied spirit, a daemon who travelled there by will-power, taking with him humans who were accustomed to flying, especially 'dried-up old crones, who since childhood have ridden over great stretches of the earth at night in tattered cloaks on goats or pitchforks'.[4] In Kepler's story, the narrator was introduced to the daemon by a wise woman who gathered herbs on the slopes of the volcano Hekla in Iceland, from which the travellers to the moon left during a lunar eclipse, voyaging in the shadow of the earth to avoid the burning rays of the sun.

This story caused Kepler big problems. When he was writing, witchcraft was taken very seriously, and it was generally believed that witches could fly like spirits; indeed it was the prevalence of this belief that made Kepler think it would be a persuasive literary device. In Kepler's home town in Germany, Leonberg, several women had just been burned as witches when news of Kepler's still unpublished book leaked out. His mother, Katherina Kepler, was accused of witchcraft, arrested and imprisoned, and Kepler had to spend several years protecting her from execution.[5]

The idea of disembodied minds soon became a central feature of mechanistic science. René Descartes in his *Meditations* (1641) took as the first principle of his philosophy, 'I am thinking, therefore I exist', and immediately inferred that his thinking mind was disembodied:

> From this I knew I was a substance whose whole essence or nature is simply to think, and which does not require any place, or depend on any material thing, in order to exist. Accordingly, this 'I' – that is, the soul by which I am what I am – is entirely distinct from the body, and indeed is easier to know than the body, and would not fail to be whatever it is, even if the body did not exist.[6]

His mind was godlike and immortal. He could know the laws of nature through his reason and participate in the mathematical mind of God. By contrast, his body was material and, like all other matter, unconscious and mechanical.

Science became a view from nowhere. The minds of scientists were somehow disembodied. This is why Stephen Hawking is such an iconic figure in the popular imagination. Through the misfortune of his illness, he is as close to a disembodied mind as a human can be. As a quote from *Time* magazine on the cover of his bestselling book, *A Brief History of Time* (1988), put it, 'Even as he sits helplessly in his wheelchair, his mind seems to soar ever more brilliantly across the vastness of space and time to unlock the secrets of the universe.' This image of disembodied minds at the same time harks back to the visionary journeys of shamans, whose spirits could travel into the underworld in an animal form, or fly into the heavens like a bird. Like the spirit of the shaman, the mind of the scientist can travel far up into the sky; he can look back from the heavens and observe the earth, the solar system, our home galaxy, and even the entire universe as if from outside. He can travel also in the other direction into the realm of the very small, zooming into the most minuscule recesses of matter.

Thought experiments have played an important part in science, most notably when Albert Einstein imagined himself running alongside a wave of light. He realised that, from the point of view of a disembodied mind travelling at the speed of light, the light would appear to be stationary and no time would elapse. This imaginary disembodied experience preoccupied him for years; he first began thinking about it in 1896 when he was sixteen, and it played a crucial role in the development of his theory of relativity.[7]

Although only exceptional scientists could use their imaginations like Kepler and Einstein, disembodied, objective knowledge was an ideal that set science apart from other forms of human knowing. In order to emphasise science's special status, scientists adopted a peculiar style of writing that became popular in the late nineteenth century and is still found in many scientific reports today. They wrote in the passive voice as if they were dispassionate, disembodied observers before whom events unfolded spontaneously. Instead of saying, 'I took a test tube,' they wrote, 'A test tube was taken.' Instead of observing, 'It was observed that . . .' Instead of someone thinking about the results, 'It was considered that . . .'

In the nineteenth century, materialists believed that physics was able to give a clear definition of matter, leaving minds out of the picture altogether, but with the development of quantum theory from the 1920s this assumption became untenable. Observations require observers, and the way in which experiments are done affects the results they give. This is obvious, but until the development of quantum theory, physicists tried to pretend that they were not involved in their own experiments. As the physicist Bernard d'Espagnat expressed it in 1976, in the latter part of the nineteenth century,

> physicists thought that they were able to define matter (as the collection of all atoms plus the fields) and believed they could formulate their science without any reference, even an implicit one, to states of consciousness of observers. Consequently, thinkers at that time legitimately believed that 'matter' thus defined was indeed the sole and primeval reality. Nowadays, however, the situation is completely different . . . [T]he principles of physics itself have undergone such an evolution that they cannot even be formulated without referring (though in some cases only implicitly) to the impressions – and thus to the minds – of the observers. As a consequence, materialism is bound to change.[8]

Nevertheless, physicists and other scientists went on using the passive voice in their reports. Things are now changing, as I discuss below, but in the popular image of science, and in much of science education, the passive voice is still employed to maintain the illusion of disembodied objectivity.

The allegory of the cave

In Plato's famous allegory of the cave, captives are chained to the wall, and see only confused shadows on the wall. They are subject to all kinds of opinion, illusion and conflict. The philosopher is like a prisoner who breaks from the cave and sees reality as it truly is.

As Bruno Latour, the sociologist of science, pointed out in his book *The Politics of Nature* (2009), this allegory took on a new lease of life in relation to science. For Plato, the allegory of the cave implied a journey beyond the realm of the body and the senses to the realm of immaterial Ideas. But its meaning has been hijacked. For materialists, objective reality is not the realm of Ideas but mathematicised matter. In the modern version of this allegory, scientists alone can step out of the cave, observe reality as it is, and come back into the cave imparting some of this knowledge to the rest of humanity, confused by rival subjectivities. Only scientists can see reality and truth. 'The Philosopher, and later the Scientist, have to free themselves from the tyranny of the social dimension, public life, politics, subjective feelings, popular agitation – in short, from the dark Cave – if they want to accede to truth.' Back within the cave, the rest of humanity is locked into the realm of multiculturalism, conflict and politics. As Latour put it,

> The allegory of the Cave makes it possible to create in one fell swoop a certain idea of science and a certain idea of the social world that will serve as a foil for science . . . The contraries turn out to be combined in one and the same heroic figure, that of the Philosopher-Scientist, at once Lawgiver and Saviour. Although the world of truth differs absolutely, not relatively, from the social world, the Scientist can go back and forth from one world to the other no matter what: the passageway closed to all others is open to him alone . . . In the original myth, as we know, the Philosopher managed only with the greatest difficulty to break the chains that attached him to the shadowy world . . . Today, sizeable budgets, vast laboratories, huge businesses, and powerful equipment allow researchers to come and go in complete safety between the social world and the world of Ideas, and from Ideas to the dark cave where they go to bring light. The narrow door has become a broad boulevard.[9]

Together with the fantasy of disembodied knowledge, the allegory of the cave implicitly supports the ideal of scientific objectivity. But the behaviour of scientists themselves is more ambiguous.

The humanity of scientists

Among the many scientists I have known, some are ruthlessly ambitious, others kind and generous; some boringly pedantic, others excitingly speculative; some narrow-minded, others visionary; some cowardly, others brave; some meticulous, others careless; some honest, others deceptive; some secretive, others open; some original, others unoriginal. In other words, they are people. They vary, just as other kinds of people vary.

Through studying scientists in action, sociologists of science have revealed that scientists are indeed like other people. They are subject to social forces and peer-group pressures, and they need acceptance, funding and, if possible, political influence. Their success does not depend simply on the ingenuity of their theories or the facts they discover. The facts do not speak for themselves. To be successful, scientists need rhetorical skills, to build up alliances and win the support of others.[10]

The historian of science Thomas Kuhn has shown that 'normal science' is practised within a shared framework of assumptions and agreed practices, a paradigm. Phenomena that do not fit – anomalies – are routinely dismissed or explained away. Scientists are often dogmatic and prejudiced when confronted with evidence or ideas that go against their beliefs. They usually ignore what they do not want to deal with. 'Turning a blind eye is the no-nonsense way to deal with potentially troublesome ideas,' observed the sociologists of science, Harry Collins and Trevor Pinch.[11] 'The meaning of an experimental result does not . . . depend only on the care with which it was designed and carried out, it depends on what people are ready to believe.'[12]

In disputes between rival scientists, experimental results are rarely decisive on their own. The facts do not speak for themselves because there is no agreement about the facts. Maybe the method

was flawed, or the apparatus faulty, or the data wrongly interpreted. When a new consensus builds up, these disputes recede
into the background, and the 'correct' results are accepted, making
it easier for similar results to be correct.

The determination of the fundamental constants is a case in
point. When the speed of light, c, apparently dropped by 20 kilometres per second from 1928 to 1945, laboratories all around the
world reported measurements close to the consensus value. But
when c went up again, laboratories duly agreed closely with the
new consensus (see Chapter 3). Did the speed of light really
change? The data say that it did. But for theoretical reasons it could
not really have changed, because it is believed to be a fundamental
constant. Therefore the consensus data must have been flawed.
The scientists probably discarded measurements that didn't fit,
and 'corrected' the remaining data until they converged on the
expected value as a result of 'intellectual phase locking' (see pages
92–3).

An international committee fixed the speed of light by definition in 1972, putting an end to embarrassing variations. But other
constants have continued to vary, especially the Universal
Gravitational Constant, G. So does G really vary? The facts cannot
speak for themselves because most of the measurements are not
published. Within individual laboratories researchers discard
unsuitable data, arriving at the final value by averaging selected
measurements. Then an international committee of experts
selects, adjusts and averages the data from different laboratories to
arrive at the internationally recognised 'best value' of G. Previous
'best values' are consigned to the archives of science, where they
gather dust.[13]

Anyone who has actually carried out scientific research knows
that data are uncertain, that much depends on the way they are
interpreted, and that all methods have their limitations. Scientists
are used to having their data and interpretations scrutinised and
criticised by anonymous peer-reviewers. They are usually well
aware of the uncertainties and limitations of knowledge in their
own field.

The illusion of objectivity gains in strength through distance. Biologists, psychologists and social scientists are notorious for physics envy, seeing physics as far more objective and precise than their own rather messy fields, where there is so much uncertainty. From the outside, metrology, the branch of physics concerned with fundamental constants, seems an oasis of certainty. But metrologists themselves make no such claim: they are preoccupied with variations in measurements, arguments about the reliability of different methods, and disputes between different laboratories. They achieve a higher level of precision than scientists studying plants, rats or minds, but their 'best values' are still consensus figures arrived at through processes of subjective evaluation.

The further the distance, the stronger the illusion. Those who are most prone to idealise the objectivity of scientists are people who know almost nothing about science, people for whom it has become a kind of religion, their hope of salvation.

The active voice

The idealised objectivity of science is reflected in the use of the passive voice in many science reports: 'A test tube was taken . . .' All research scientists know that writing in the passive voice is artificial; they are not disembodied observers, but people doing research. Technocrats also use the passive voice to give their reports an air of scientific authority, dressing up opinions as objective facts.

The passive style did not become fashionable in science until the end of the nineteenth century. Previously, scientists like Isaac Newton, Michael Faraday and Charles Darwin used the active voice. The passive was introduced to make science seem more objective, impersonal and professional. Its heyday in the scientific literature was from 1920 to 1970. But times are changing. Many scientists abandoned this convention in the 1970s and 1980s.

In 1999, I was astonished to read in my eleven-year-old son's science notebook, 'The test tube was heated and carefully smelt.'

At primary school his science reports had been lively and vivid, but when he moved to secondary school they became stilted and artificial. His teachers had told him to write that way, and gave him a style sheet to copy.

I'd thought that schools had abandoned this practice years ago, and was curious to find out how widespread it still was. In 2000, I carried out a survey of 172 secondary schools in Britain to discover how many insisted on the passive style. Overall, 42 per cent of the schools still promoted the passive voice, 45 per cent the active, and 13 per cent had no preference.[14]

Most of the teachers enforcing the use of passive voice said they were simply following convention. No one was enthusiastic about it. They taught it out of a sense of duty because they believed that leading scientists and journals required it. Some thought that examination boards insisted on it, but this was not true. I found that all the UK examination boards accepted reports in the active or the passive voice.[15]

I also found that most scientific journals accepted papers in the active voice; some, including *Nature*, positively encouraged it. I surveyed fifty-five journals in the physical and biological sciences, and found only two that required passive constructions.

When Lord May, the president of the Royal Society, read the results of my survey of school science teaching, he was 'horrified' that so many favoured the passive: 'I would put my own view so strongly as to say that, these days, the use of the passive voice in a research paper is the hallmark of second-rate work,' he said. 'In the long run, more authority is conferred by the direct approach than by the pedantic pretence that some impersonal force is performing the research.'[16] May's views were shared by many other eminent scientists, including the Astronomer Royal, Martin Rees, who succeeded Lord May as president of the Royal Society, and Bruce Alberts, then president of the US National Academy of Sciences.

Nevertheless, old habits die hard, and science teachers in many schools still insist that their pupils write in the passive voice. In a survey in 2010, science teachers in 30 per cent of British secondary schools were still insisting on the passive voice.[17] This is an outdated practice. 'Primary and secondary teachers should, without any reservation, be encouraging all their students to be writing in the active voice,' said Lord May.[18]

Switching from the passive to the active voice in science reports is a simple reform that costs nothing and makes science writing more truthful and more readable.

Standard scientific pretences

Peter Medawar was an articulate British biologist who won the Nobel Prize for Medicine. In a witty talk on BBC radio in 1963, he asked, 'Is the scientific paper a fraud?' and answered, 'Yes.' He was not referring to fraudulent data, but to the way that scientific papers are conventionally written. In scientific journals, the standard format for articles, then as now, is to start with a neutral-sounding Introduction, setting out the problem and referring to earlier research, followed by a Methods section, then the Results and finally a Discussion. As Medawar described it,

> The section called 'results' consists of a stream of factual information in which it is considered extremely bad form to discuss the significance of results you are getting. You have to pretend firmly that your mind is, so to speak, a virgin receptacle, an empty vessel, for information which flows into it from the external world for no reason which you yourself have revealed. You reserve all appraisal of the scientific evidence until the 'discussion' section, and in the discussion you adopt the ludicrous pretence of asking yourself if the information you have collected actually means anything.

Medawar pointed out that this procedure, still standard today, gives an entirely false impression of the way science works, suggesting that scientists collect facts and then draw general conclusions from them. In fact, scientists start with an expectation or a hypothesis that provides the incentive for the enquiry in the first place. It is only in the light of these expectations that some observations are thought to be relevant, and others not; that some methods are chosen and others discarded; that some experiments are done rather than others. Medawar suggested a more honest approach: putting the discussion at the beginning:

> The scientific facts and scientific acts should follow the discussion, and scientists should not be ashamed to admit, as many of them apparently are ashamed to admit, that hypotheses appear in their minds along uncharted by-ways of thought; that they are imaginative and inspirational in character; that they are indeed adventures of the mind.[19]

How experimenters affect their results

Most medical researchers are well aware that their beliefs and expectations can influence the results of their experiments, which is why many clinical trials are carried out double-blind: neither the researchers nor the patients know who has received which treatment (see Chapter 10).

Experimenter effects are also well known in experimental psychology. This principle was illustrated in a classic experiment in which the experimenters trained a group of psychology graduate students to administer the Rorschach test, in which subjects were asked to identify patterns in inkblots. The experimenters told half of the students that experienced psychologists obtained more human than animal images from their subjects. They told the other half of the group the opposite. Sure enough, when they administered the test the second group found more animal images than the first.[20]

Even in experiments with animals, experimenters' expectations can influence the results. In a classic experiment at Harvard, Robert Rosenthal and his colleagues instructed students to test rats in standard mazes. They asked them to compare two strains of rat produced by generations of selective breeding for good and poor performance in mazes. But they deliberately deceived their students. In fact, the rats came from a standard laboratory strain and were divided at random into two groups labelled 'maze-bright' and 'maze-dull'.

Trusting what they had been told, the students expected the bright rats to do better than the dull ones and, sure enough, they found that the 'bright' rats learned much faster than the 'dull' ones.[21] Since the rats were more or less identical, these dramatic differences must have resulted from the students' expectations.

Although experimenter-expectancy effects are widely recognised in psychology and medicine, in the 'hard' sciences most scientists assume that they are irrelevant. They take it for granted that their own expectations have no influence on their experiments and on the recording of data.

From 1996 to 1998, I carried out a survey of more than 1,500 papers in leading scientific journals to find out how often the researchers used blind methods. Caroline Watt and Marleen Nagtegaal later replicated this survey, using a different selection of journals (Table 1).

TABLE 11.1. A comparison of the percentage of papers reporting blind methodologies in different fields of science in two independent surveys, by Sheldrake (1999c), and Watt and Nagtegaal (2004).

Research field	Per cent blind, 1999[22]	Per cent blind, 2004[23]
Physical sciences	0	0.5
Biological sciences	0.8	2.4
Animal behaviour	2.8	9.3
Psychology	7.0	22.5
Medical sciences	24.2	36.8
Parapsychology	85.2	79.1

Watt and Nagtegaal found a higher percentage of papers with blind methodologies in most areas than I did, and a slightly lower percentage in parapsychology, but in both our surveys, in the physical sciences almost no research involved blind methodologies, and in the biological sciences very little, less than 2.5 per cent.

Even in experimental psychology, animal behaviour and the medical sciences, where the effects of experimenters' expectations are widely recognised, a minority of studies used blind methods. By far the highest percentage was in parapsychology.

I also organised a telephone survey of senior researchers in fifty-five departments in eleven British universities, including Oxford, Cambridge, Edinburgh and Imperial College, London. My research assistant, Jane Turney, carried out the interviews by telephone. She asked the professors or other senior scientists if anyone in their department used blind methodologies, and also whether they taught students about such methods.

Some of the scientists did not know what was meant by the phrase 'blind methodology'. Most were aware of blind techniques, but said that they were necessary only in clinical research or psychology. They thought they were used to avoid biases introduced by human subjects. The commonest view among physical and biological scientists was that blind methodologies were unnecessary because 'Nature itself is blind', as one researcher put it. A professor of chemistry added, 'Science is difficult enough as it is without making it even harder by not knowing what you are working on.'

Out of twenty-three physics and chemistry departments, only one used blind methods and taught the students about such methods. Out of forty-two departments in the biological sciences, twelve (29 per cent) sometimes used blind methods and taught about them.[24] But only in exceptional cases were blind techniques used routinely. My survey revealed three examples, all of which involved industrial contracts that required the university scientists to evaluate coded samples without knowing their identity.[25]

Experimental tests for experimenter effects

The assumption that blind techniques are unnecessary in most fields of science is so fundamental that it deserves to be tested.[26] In all branches of experimental science we can ask: can the expectations of experimenters act like self-fulfilling prophecies, introducing a bias, conscious or unconscious, into the way the data are collected, analysed and interpreted?

There is a simple way to find out by doing experiments on experiments. Take a typical experiment involving a test sample and a control; for example, the comparison of an inhibited enzyme with an uninhibited control enzyme in a biochemical experiment. Then carry out the experiment as usual, where the experimenter knows which sample is which. Also do the experiment under 'blind' conditions with the samples labelled A and B. In a student practical class, for instance, half of the class would do the experiment blind, while the other half would know which sample is which. If there were no significant differences between the results under blind and open conditions, this would show that blind techniques were unnecessary. Significant differences would reveal the existence of experimenter effects. Further research would then be needed to find out how these effects worked.

This experiment costs nothing but simply involves labelling samples differently. It would be easy to do in laboratory classes in schools or universities. When I first proposed this simple experiment,[27] I naïvely assumed that skeptics, who spend so much of their time insisting on the objectivity of science, would be particularly interested in this question. I therefore launched an appeal in the *Skeptical Inquirer*[28] and in the *Skeptic*,[29] asking people who worked in universities to collaborate in this research. There was no response. Richard Wiseman, himself a skeptic, together with Caroline Watt, launched another appeal in the *Skeptical Inquirer*[30] with a similar lack of response.

On one occasion I thought that it was going to be possible to do this test when a physics teacher at one of Britain's leading schools agreed to try it with his final-year students. But he had to ask permission from the head of science, who asked me to meet him to explain what I had in mind. His response was illuminating. He said, 'Of course the students are going to be influenced by their expectations. That's what science education is all about. It's obvious they will try to get the right results. This experiment will open up a can of worms, and I don't want it opened in my school.'

These remarks were helpful because of their directness and honesty. I realised that all professional scientists have spent years doing lab classes at school and at university being trained to get the expected results.

Over a period of ten years at Cambridge University (in cell biology and biochemistry), and one year at Harvard (in Biology 101), I taught in laboratory classes in which students did standard experiments with outcomes that were well known in advance. But there were always some students who did not get the 'right' results. Everyone assumed they had simply made mistakes. Some students were often bad at getting the standard results: I suppose they graduated with poor degrees, and were therefore unlikely to go on to a career in scientific research. Those who became professional scientists were people who showed a reliable ability to get the correct results over many years of practical education in laboratories.

Although experimenter effects may often result from biases in the observation and recording of results, experimenters might affect the experimental system itself. This is easy to understand when experiments involve human subjects, who may well respond to the experimenters' expectations and attitude. Rosenthal's classic experiment with Harvard students testing rats shows that animals too can be influenced by the way they are treated. But there is a more radical possibility. In the uncertain circumstances of research, the experimenter's expectations may directly affect the system under investigation through mind-over-matter effects or psychokinesis. For example, if hundreds of highly qualified physicists expect to find an evanescent particle among the indeterminate events that occur in a particle accelerator, could their expectations affect these quantum events? Could the hopes of scientists influence the outcomes of more mundane experiments too?

These may seem far-fetched possibilities, and discussing them is normally prevented by the taboo against psychic phenomena. But I believe it is important to investigate rather than suppress this question. Many stories circulate in laboratories that suggest that some people bring about mysterious effects. Sometimes they are negative effects, or jinxes. One of the most famous examples is the

so-called Pauli effect, named after the Nobel-Prize-winning physicist Wolfgang Pauli (1900–58). He was reputed to cause the failure
of laboratory equipment merely by his presence. For fear of this
effect, his friend Otto Stern, an experimental physicist, banned
Pauli from his laboratory in Hamburg. Pauli himself was convinced
that the effect was real, and was worried that he might have
contributed unwittingly to the burning of the cyclotron at
Princeton University when he was nearby.[31]

Sometimes apparent mind-over-matter effects are positive. A
professor of biochemistry from a major US university told me that
part of the secret of his success was that he could achieve better
purifications of protein molecules than his colleagues. He said
that when a sample of mixed proteins was being separated, he
stayed with the apparatus in the cold room 'willing' the system to
give clearer separations, and saying, 'Separate!'

Was this a personal superstition, or did it have any effect? This
question could be investigated experimentally. For example, two
identical pieces of apparatus could be loaded with the same
mixture of proteins. One, selected at random, would then be given
to the professor to accompany during the separation process. The
other would be put in a different cold room and left alone for the
same period of time. The separations would then be compared to
see if there was any difference. I tried to persuade this professor to
do the experiment himself, but he was unwilling to try it. Although
he was curious, he could not risk the potential damage to his credibility and career.

The supposed objectivity of the 'hard sciences' is an untested
hypothesis. There is a conspiracy of science about experimenter-
expectancy effects in most branches of physics, chemistry and
biology. The assumption that they are confined to clinical research,
human psychology and animal behaviour may well be untrue.

Another problem is that scientists usually publish only a small
proportion of their data. If they cherry-pick the results that suit
their hypotheses, this will introduce another source of bias, sometimes called 'publication bias', and sometimes called the 'file-drawer
effect', because negative results are left in files (see Chapter 9).

The built-in bias of publication

Of all areas of scientific research, parapsychology is subject to the most severe and persistent skeptical scrutiny, as discussed in Chapter 9. Skeptics are strongly motivated to dismiss any positive findings, and have a ready-made list of objections: flawed methods, fraud, experimenter effects or the selective publication of positive results. Because parapsychologists are so aware of these standard criticisms, they are exceptionally careful to carry out their experiments as rigorously as possible. In the surveys summarised in Table 11.1, a far higher proportion of parapsychologists used blind methods than researchers in any other branch of science. Parapsychologists are also far more rigorous about publishing negative results and controlling against the so-called file-drawer effect.[32]

Skeptics are right to point out these possible sources of error in parapsychological research, and their continual scrutiny has benefited the standards of research in the subject. But the same sceptical principles should be applied to other areas of science. What proportion of research results are published in physics, chemistry and biology? There seem to have been no studies on this question, but in informal surveys I have carried out myself, in most subjects this proportion seems to be around 5 to 10 per cent.

Scientists are more likely to publish their 'best' results than negative or inconclusive findings. We saw one example in Chapter 10: the manufacturers of Prozac, Eli Lilly, published positive but not negative results from clinical trials. In addition, scientific journals are often unwilling to publish negative results. The implications are enormous. As Ben Goldacre put it, 'whole fields of science are at risk of spurious positive findings'.[33]

Published data have to pass through three selective filters. The first filtration of the data occurs when experimenters decide to publish some results rather than others; the second when editors of journals consider only certain kinds of results eligible for publication; and the third in the peer-review process, which ensures that expected results are more likely to be approved for publication than unexpected results.

If businesses were required to publish only 10 per cent of their accounts, they would probably publish those that made their business look as profitable and as well managed as possible. Conversely, if they needed to submit only 10 per cent of their accounts to the tax authorities, they would tend to show their least profitable activities. Suppressing 90 per cent of the data gives a lot of scope for selective reporting. How much does this practice affect the sciences? No one knows.

Scientific fraud and deceit

Scientists, like doctors, lawyers and other professionals, generally resist attempts by outsiders to regulate their conduct. They pride themselves on their own system of controls, which are threefold:

1. Applications for jobs and grants are subject to peer review, ensuring that the researchers and their projects meet the approval of established professionals in their field.
2. Papers submitted to scientific journals are peer-reviewed, and have to pass the critical scrutiny of expert referees, usually anonymous.
3. All published results are potentially subject to independent replication.

Peer review and refereeing procedures can indeed act as important quality checks, and are often effective, but they tend to favour expected results and conventional procedures. Independent replications are rarely performed. There is usually no motivation for repeating the work of others. And even if exact replications are performed, it is difficult to get them published because scientific journals favour original research. Generally speaking, scientists try to replicate other people's results only when the results are of unusual importance or when fraud is suspected on other grounds.

An additional safeguard is the convention that when other scientists ask to see a researcher's raw data so they can reanalyse them, the data are supplied, in the interests of openness. However, when I asked for data from scientists making sceptical claims in fields of research closely related to my own, they have refused to supply them, either on the grounds that they were 'inaccessible' or because they planned to reanalyse them themselves (but never did). In a recent systematic study, some Dutch psychologists at the University of Amsterdam contacted the authors of 141 papers published in leading psychology journals, asking for access to the raw data for the sake of reanalysis. All these journals required authors to sign an undertaking that they 'would not withhold the data on which their conclusions are based from other competent professionals'. After six months and four hundred emails, the Amsterdam researchers received sets of data from only 29 per cent of the authors.[34]

One of the few areas of science under a limited form of external supervision is the testing for safety of new foods, drugs and pesticides. In the United States, every year many thousands of results are submitted by industry for review by the Food and Drug Administration (FDA) or the Environmental Protection Agency (EPA). Their inspectors continually unearth falsified data.[35]

Frauds in the unpoliced hinterlands of science are rarely exposed by the official mechanisms of peer review, refereeing or independent replication. Most come to light as a result of whistle-blowing by colleagues or rivals, often as a result of a personal grievance. When this happens, the typical response of the authorities is to try to hush the matter up. If the charges of fraud do not blow over and if the evidence becomes overwhelming, then an official inquiry is held, and someone is found guilty and dismissed in disgrace.[36]

Probably many cases of fraud are indeed hushed up. The authorities have a strong motive not only to protect the reputation of their institution but the image of science itself. The philosopher Daniel Dennett argues that beliefs are social forces in their own right, and that a *belief in belief* plays a vital role in sustaining social institutions. Some beliefs need to be maintained for the general good. For instance, democracy depends on maintaining a belief in democracy. Likewise, the authority of science depends on maintaining a belief in scientific authority: 'Since the belief in the integrity of scientific procedures is almost as important as the actual integrity, there is always a tension between a whistle-blower and the authorities, even when they know that they have mistakenly conferred scientific respectability on a fraudulently obtained result.'[37]

One of the biggest cases of fraud to be exposed in physics in the twenty-first century concerned Jan Hendrik Schön, a young researcher on nanotechnology at Bell Laboratories, in New Jersey. He seemed brilliantly successful and amazingly productive, making breakthrough after breakthrough and receiving three prestigious awards. But in 2002, several physicists noticed that the same data appeared in different papers, apparently from different experiments. An investigating committee found sixteen instances of scientific misconduct, mostly the making up or recycling of data. As a result of the inquiry, twenty-eight papers were withdrawn by scientific journals, including nine in *Science* and seven in *Nature*.[38] Schön's co-authors were declared to be innocent, although they had shared in the credit when the results were thought to be genuine. Significantly, none of these instances of fraud was detected in the peer-review process.

In another recent case, Marc Hauser, a Harvard professor of biology, was found guilty of scientific misconduct by an official inquiry at Harvard in 2010. He had falsified or invented data in experiments on monkeys.[39] Again, his dishonesty was not detected by peer reviewers, but came to light when a graduate student blew the whistle. Hauser is the author of a book called *Moral Minds: The Nature of Right and Wrong* (2007), in which he claims that morality is an inherited instinct, produced by evolution and independent of religion. Hauser is an atheist, and claims his findings support an atheist point of view. In an interview a few months before his fraud was exposed, he said that his research showed that 'atheists are just as ethical as churchgoers'.[40]

In an insightful study of fraud and deceit in science, William Broad and Nicholas Wade showed that deceptions easily pass unchallenged as long as the results are in accordance with prevalent expectations:

Acceptance of fraudulent results is the other side of that familiar coin, resistance to new ideas. Fraudulent results are likely to be accepted in science if they are plausibly presented, if they conform with prevailing prejudices and expectations, and if they come from a suitably qualified scientist affiliated with an elite institution. It is for the lack of all these qualities that new ideas in science are likely to be resisted . . . For the ideologists of science, fraud is taboo, a scandal whose significance must be ritually denied on every occasion.[41]

Scientists usually assume that fraud is rare and unimportant because science is self-correcting. Ironically, this complacent belief produces an environment in which deception can flourish.[42]

Scepticism as a weapon

Research scientists, aware of the limitations and ambiguities of their work, rarely claim to have achieved certainty, and they are routinely subject to peer review. Scepticism is an essential part of science. But it can easily be turned into a weapon to attack opponents. For example, creationists, who deny evolution, use the techniques of critical thinking to highlight problems with evolutionary theory and expose weaknesses in the evidence, such as gaps in the fossil record. Is this because they are seeking truth? No. They believe they already know the truth. Scepticism is a weapon to defend their beliefs by attacking their opponents.

The same techniques have been used for years by organised groups of skeptics to attack psychic research, parapsychology and alternative medicine. Their motives are primarily ideological: they, too, believe they know the truth already – psychic phenomena are illusory and mechanistic medicine is the only kind that really works (see Chapters 9 and 10).

Scepticism is also an important weapon in the defence of commercial self-interest. The publication in 1964 of the US Surgeon General's report *Smoking and Health*, based on a review of more than seven thousand scientific studies, made it clear that smoking caused lung cancer and increased the risk of suffering from emphysema (caused by the destruction of lung tissue), bronchitis and heart disease. The tobacco industry responded by setting up the Council for Tobacco Research, which funded projects at more than a hundred hospitals, universities and research labs. Many of these studies looked for complicating factors that would muddy the waters. As an executive of the cigarette company Brown and Williamson put it in 1969, 'Doubt is our product since it is the best means of competing with the "body of fact" that exists in the mind of the general public.'

By the late 1970s the tobacco industry was facing scores of lawsuits in the United States claiming personal injury from smoking. In 1979, Colin Stokes, the former chairman of the R. J. Reynolds tobacco company, addressed a meeting of tobacco-company executives to report on progress. The attacks on smoking, he told his audience, were based on studies that were either 'incomplete . . . or relied on dubious methods or hypotheses and faulty interpretations'. Tobacco-industry-funded research would supply new hypotheses and interpretations to 'develop a strong body of scientific data or opinion in defense of the product'. Above all, it would supply expert witnesses who could testify in courts.

This strategy had worked in the past, and there was no reason to think it would not work in the future. Stokes boasted, 'Due to favorable scientific testimony, no plaintiff has ever collected a penny from any tobacco company in lawsuits claiming that smoking causes lung cancer or cardiovascular illness.'[43] In the end Stokes's strategy failed, but it staved off legal cases and delayed anti-smoking legislation for years.

The tobacco-industry strategy was adopted by numerous other industries defending toxic chemicals such as lead, mercury, vinyl chloride, chromium, benzene, nickel and many more. David Michaels, who was assistant secretary for environment, safety and health at the US Department of Energy in the late 1990s, saw first-hand how corporate interests worked to defeat the regulation of beryllium, a chemical element originally used to increase the yield of nuclear explosions, and later used in the manufacture of electronic and other consumer items. Following the discovery in the 1940s that beryllium can scar lung tissue, the Atomic Energy Commission established a safe level of exposure at two micrograms per cubic metre of air. By the 1990s it was clear that people were falling sick at levels far lower. When the federal government began the process of revising exposure limits, the leading US producer of beryllium, Brush Wellman, fired back with a series of reports suggesting that the physical properties of beryllium particles might influence its toxicity. Thus no action should be taken until these factors could be worked out more precisely. By 'manufacturing uncertainty', Brush Wellman staved off life-saving regulations.[44]

Emphasising uncertainty on behalf of big business has become a big business in itself. Specialised product-defence firms have increasingly skewed the scientific literature, created and magnified scientific uncertainty, and influenced policy decisions to the advantage of polluters and the manufacturers of dangerous products. In fact, the science behind any proposed public-health or environmental regulation is now almost always challenged, however powerful the evidence.

The strategy of dismissing research conducted by mainstream scientists as 'junk science' and elevating science conducted by product-defence specialists as 'sound science' creates confusion and undermines the public's confidence in science's ability to address public-health and environmental concerns.[45]

All these issues took on a global significance in relation to climate change. Organised attempts to discredit the growing scientific consensus began in 1989, with a report attacking climate science by the George C. Marshall Institute, which was originally established in 1984 to defend President Reagan's Strategic Defense Initiative ('Star Wars') against attacks by other scientists. The Marshall report blamed global warming on increased activity by the sun, discounting the effect of greenhouse gases. This is not the place to review the ongoing controversies, but the Marshall Institute and oil-industry-funded scientists have continually muddied the waters of the debate.[46]

In practice, the goal of scepticism is not the discovery of truth, but the exposure of other people's errors. It plays an essential role in science, religion, scholarship, business, journalism, politics, the legal system and common sense. But we need to remember that it is often a weapon serving belief or self-interest.

Facts and values

The illusion of scientific objectivity sustains the equally illusory distinction between facts and values, on which institutional science has been based from the outset. Francis Bacon (1561–1626) made a distinction between the innocent knowledge of nature, given by God to Adam before the Fall, and the knowledge of good and evil, or values, that caused the Fall (see page 13). But Bacon was disingenuous. He also coined the slogan 'knowledge is power', which has been the basis on which scientists have solicited funds for their research from governments and commercial companies ever since. Very few patrons of scientific research are interested in innocent knowledge for its own sake. When scientists submit grant proposals, they almost always claim that their research will be useful. The facts they hope to discover will be of value for national defence, combating disease, increasing profits, promoting the yields of crops, improving navigation, increasing national prestige, or conferring other benefits. The hoped-for values come before the facts; the promised values enable the research to be funded and the facts to be established.[47]

As we have seen in this chapter, facts and values are not clearly separated, and scientists are all too human. Nevertheless, scientists have found out far more than anyone knew before, and the sciences have transformed the conditions of human life. But the myths and ideology on which they are based have become unconscious habits of thought, creating unhelpful illusions that imprison scientific enquiry and fuelling prejudice and dogmatism. In the final chapter, I suggest that the best way forward is to recognise the plurality of sciences, natures and points of view.

Questions for materialists

Experimenters' expectations are known to affect the results of research in psychology, parapsychology and medicine, which is why researchers often use blind methodologies. Do you think that experimenter effects could play a role in other fields of science too?

Do you think that scientists and science students should write in the passive voice in their reports, or use the active voice?

Most scientists publish only a small proportion of their results. Do you think that this is likely to introduce serious biases into the scientific literature?

How should scientists deal with ideologically, politically or commercially motivated scepticism?

SUMMARY

Scientists are often imagined to achieve a superhuman level of objectivity. This belief is sustained by the ideal of disembodied knowledge, unaffected by ambitions, hopes, fears and other emotions. In the allegory of the cave, scientists venture forth into the light of objective truth and bring back their discoveries for the benefit of ordinary people, trapped in a world of opinion, self-interest and illusion. By writing in the passive voice ('a test tube was taken') rather than the active voice ('I took a test tube') scientists tried to emphasise their objectivity, but many have now abandoned this pretence. Scientists are, of course, people, and subject to the limitations of personality, politics, peer-group pressures, fashion and the need for funding. Within medicine, psychology and parapsychology, most researchers recognise that their expectations can bias their results, which is why they often use blind or double-blind methodologies. In the so-called hard sciences, most researchers assume that blind methods are unnecessary. This is no more than an assumption, and needs to be tested experimentally. In most fields of science, researchers publish only a small proportion of their data, giving plenty of scope for the selective presentation of results, and scientific journals introduce a further source of bias through their unwillingness to publish negative findings. Fraud and deceit in science are rarely detected by the peer-review system and usually come to light as a result of whistle-blowing. Scepticism is a healthy part of normal science but is often used as a weapon in defence of politically or ideologically motivated points of view, or to stave off the regulation of toxic chemicals. Product-defence companies emphasise uncertainty on behalf of big business, influencing policy decisions in favour of their clients. The separation of facts and values is usually impossible in practice, and many scientists have to exaggerate the value of their research in order to get it funded. Although the objectivity of science is a noble ideal, there is more hope of achieving it by recognising the humanity of scientists and their limitations than by pretending that science has a unique access to truth.

12

Scientific Futures

The sciences are entering a new phase. The materialist ideology that has ruled them since the nineteenth century is out of date. All ten of its essential doctrines have been superseded. The authoritarian structure of the sciences, the illusions of objectivity and the fantasies of omniscience have all outlived their usefulness.

The sciences will have to change for another reason too: they are now global. Mechanistic science and the materialist ideology grew up in Europe, and were strongly influenced by the religious disputes that obsessed Europeans from the seventeenth century onwards. But these preoccupations are alien to cultures and traditions in many other parts of the world.

In 2011, the worldwide expenditure on scientific and technological research and development was more than $1,000 billion, of which China spent $100 billion.[1] Asian countries, especially China and India, now produce enormous numbers of science and engineering graduates. In 2007, at BSc level there were 2.5 million science and engineering graduates in India and 1.5 million in China,[2] compared with 515,000 in the United States[3] and 100,000 in the UK.[4] In addition, many of those studying in the United States and Europe are from other countries: in 2007, nearly a third of the graduate students in science and engineering in the United States were foreign, with the majority from India, China and Korea.[5]

Yet the sciences as taught in Asia, Africa, the Islamic countries and elsewhere are still packaged in an ideology shaped by their European past. Materialism gains its persuasive power from the technological applications of science. But the successes of these applications do not prove that this ideology is true. Penicillin will go on killing bacteria, jet planes will keep on flying and mobile telephones will still work if scientists move on to wider views of nature.

No one can foresee how the sciences will evolve, but I believe recognising that 'science' is not one thing will facilitate their development. 'Science' has given way to 'the sciences'. By moving beyond physicalism, the status of physics has changed. By freeing the sciences from the ideology of materialism, new opportunities for debate and dialogue open up, and so do new possibilities for research.

From one science to many sciences

Mechanistic science appeared to provide a simple, unified view of nature. Everything was made up of ultimate particles of matter whose properties and movements were governed by eternal mathematical laws. Theoretical physicists are still striving for a Theory of Everything and hope a unified formula will explain all of reality in terms of the properties of subatomic particles and the forces that affect them (see Chapter 1). Everything can ultimately be reduced to physics. As Lee Smolin expressed the conventional view, 'Twelve particles and four forces are all we need to explain everything in the known world.'[6]

This naïve, old-fashioned reductionist faith bears no relation to the reality of the sciences. Physiologists do not explain blood pressure in terms of subatomic particles but through the pumping activity of the heart, the elasticity of arterial walls, and so on. Linguists do not analyse languages in terms of the movements of subatomic particles in the molecules in the air through which the sounds of voices travel: they study the patterns of words, grammars and meanings. Botanists do not study the evolution of flowers by probing the atoms within them, but by comparing their structures and relationships to living and extinct species. As the physicist John Ziman put it,

At successively higher levels of complexity, from elementary particles and chemical molecules, through unicellular and multicellular organisms, to self-aware human beings and their cultural institutions, we find systems obeying entirely novel principles. The behaviour of such systems is not predictable from the properties of their constituents, so distinct 'languages' are required to describe them scientifically. The plurality of our sciences is thus an irreducible feature of the universe we live in.[7]

There are many sciences and many natures. There is no one 'scientific method'; different sciences use different methods.[8] Geologists studying rocks make different kinds of observations from astronomers investigating distant galaxies with radio telescopes, or from biochemists studying the properties of protein molecules, or from ecologists studying rainforests. Some sciences involve experiments. Others do not. An astronomer cannot manipulate a star to see how it responds, and a paleontologist cannot travel back in time to change the way sediments formed in the ocean aeons ago. Some kinds of science are highly mathematical, like theoretical physics; others are not, like the taxonomy of dragonflies.

'Science' is an abstraction. Scientists work within specialised disciplines, and students study one or more of the sciences. At university, they have to choose between a wide range of possibilities. For example, at Cambridge University in 2011, a second-year student of natural sciences had to take three courses from the following list:[9]

animal biology
biochemistry and molecular biology
cell and developmental biology
chemistry A (mainly theoretical)
chemistry B (inorganic, organic and biological)
ecology
experimental psychology
geological sciences A (surface environments)
geological sciences B (subsurface processes)
history and philosophy of science
materials science

mathematics
neurobiology
pathology
pharmacology
physics A (mainly quantum physics)
physics B (mainly mechanics, electromagnetism and thermo-
    dynamics)
physiology
plant and microbial sciences

Each of these courses is broadly based and covers a range of specialities; for example, in animal biology, there are sections on
ecology, brains and behaviour, insect biology, vertebrate evolutionary biology and evolutionary principles. No one studies
'science', and fewer than 20 per cent study history and philosophy
of science.

Students absorb their general views about the nature of reality
as implicit assumptions or from the writings of scientific popu-
larisers. The doctrines of materialism are not taught explicitly,
and many students and scientists are unaware of their influence
in shaping the practice and assumptions of their field. For
example, most neuroscientists take it for granted that minds are
in brains and that memories are stored as material traces. These
assumptions are not treated as aspects of a philosophy of nature,
or as hypotheses to be tested: they are part of the standard paradigm or consensus reality, protected by taboos against deviant
thinking.

Ironically, the fragmentation of the sciences into separate disciplines was the stimulus for coining the word 'scientist'. At the third
annual meeting of the British Association for the Advancement of
Science in 1833, delegates expressed the need for an umbrella term
to cover their diverse interests, and William Whewell, a mathematical astronomer, suggested 'scientist'. The term was an
immediate success in America. In Britain, where scientific research
was still for the most part an expensive occupation for the leisured
classes, 'scientist' was slow to displace older terms like 'man of

science', 'naturalist' or 'experimental philosopher'. But as research increased and education expanded, there were more opportunities for employment and scientists gradually became paid professionals.[10]

As the sciences grew in power and prestige, so did the need to assert their status and authority. Patricia Fara, a historian of science, summarised the situation in the nineteenth century thus:

> Hungry for prestige, scientists wanted the authority to declare that they were incontrovertibly right, that the knowledge they produced in their laboratories was irrefutably correct. New specializations were being invented, but not all of them were deemed worthy to be labeled science. Science was splintering into disciplines – but disciplining meant controlling as well as teaching. Like police guards patrolling national borders, scientists decreed which topics should be inside the large domain they ruled over, and which should be outlawed.[11]

There are now hundreds of scientific specialities, all with their own professional societies, journals and conferences. Specialists are famously said to know more and more about less and less, and in the sciences this process has continued to produce ever more fragmented fields of knowledge, all with their own specialised publications. By 2011, there were about twenty-five thousand scientific journals.[12]

It is not the job of all these specialists to think about the underlying philosophical assumptions of the sciences. Historians and philosophers of science think about them, but they themselves are in a specialised field, often treated as of marginal interest to the real business of science. By default, the old materialist or physicalist ideology persists almost unquestioned. One of its effects is to put physics at the top of the scientific hierarchy, because physicalism by definition states that everything is ultimately explicable in terms of physics.

Physicalism and physics

Physics is the source of the vision of a simple, unified view of nature, and physicists like to think that their discipline is the most fundamental, unifying all the sciences. It is true that all material bodies are made up of quantum particles, that all physical processes involve flows of energy and all physical events happen within the framework of space-time given by the universal gravitational field. But these aspects of physics leave out almost all the details we might want to know about the growth of pine trees, the effects of sex hormones, the social life of bees, the evolution of Indo-European languages, or the design of computer software.

Ironically, for those who would like to reduce everything to physics in the interests of unifying nature, physics itself has resisted unification for decades. Its two most fundamental theories, quantum mechanics and general relativity theory, are incompatible. General relativity applies to the large-scale structure of the universe – planets, stars and galaxies – and describes gravitation, one of the four 'fundamental forces'. Quantum mechanics describes the other three forces (electromagnetism and the strong and weak nuclear forces) and is most accurate at the atomic and subatomic scales. But the two theories start from different assumptions, and have resisted years of efforts to unify them.[13]

This is where superstring and M-theories come in, with ten and eleven dimensions respectively (see Chapter 3). But instead of giving a new unity to physics, they generate vast numbers of possible worlds. The price of unification is a runaway proliferation of universes. All except our own are unobserved and unobservable. What kind of unification is this? It looks more like the ultimate plurality.

In mechanistic science, physics came first historically, growing out of the study of mechanics, astronomy and optics in medieval universities. Physics also comes first in term of prestige because of its claim to deal with the most fundamental realities as well as the origin of all things in the Big Bang. But this priority is arbitrary.

Other professional groups could claim that the status of their field is as high if not higher. Consciousness studies could claim primacy because physics happens in human minds and entirely depends on human consciousness. Maxwell's equations and superstring theories do not exist 'out there' as independent facts: they are mental constructs.

Brain scientists then could claim that without neurophysiology and brain chemistry there could be no human consciousness. Proponents of linguistics could argue that without language there would be no human culture; social scientists could claim that without societies no physics could ever have happened; economists could claim that without a functioning economy no one would be able to do physics. Meanwhile physiologists could point out that the brain is simply one part of the body, and is dependent on the co-ordinated function of the whole, including digestion, breathing, circulation, limbs, sense organs and so forth. Embryologists could argue that without embryological development there would be no bodies and no physiology to start with, and hence no physicists, and geneticists could argue that without genes there would be no embryology.

Evolutionists could point to the evolutionary origins of humanity; ecologists could stress the interdependence of all life; plant scientists could emphasise that humans and all other animals ultimately depend on plants for food, and on the biochemistry of photosynthesis; then physicists could re-enter the picture with solar physics and astronomy, without which there would be no photosynthesis. Engineers and technologists could argue that without scientific apparatus no accurate measurements would be possible, and without modern communications technologies and computers the sciences would not be able to function. And so on.

No one can claim absolute primacy. Everything is interlinked. Nothing is permanent and isolated from everything else. There is an interdependence of all things and all levels of organisation. This sounds very like the Buddhist doctrine of dependent origination or dependent arising, according to which all phenomena occur in a mutually interdependent web of cause and effect.

The materialist philosophy and the primacy of physics go hand in hand. So do the interdependence of all realities and the plurality of the sciences. The sciences still need unifying principles, but they need not come exclusively from physics.

Unifying principles

As well as the familiar unifying principles of physics, like forces, fields, and flows of energy there is the principle of organisation in nested hierarchies. Systems, or organisms, or holons, or morphic units at every level, are wholes made up of parts, which in turn are wholes made up of parts. Crystals contain molecules, which contain atoms, which contain subatomic particles. Galactic clusters contain galaxies, which contain solar systems, which contain planets. Societies of organisms contain animals, which contain organs, which contain tissues, which contain cells, which contain molecules, which contain atoms . . . (see Chapter 1).

The hypothesis of morphic resonance provides another unifying principle: all self-organising systems draw upon a collective memory from similar systems of their kind (see Chapters 3, 6 and 7).

But whenever we find general principles, their very generality hides the details of specific things. Sequoias, seaweeds and sunflowers all consist of the same chemical elements, capture the energy of light by photosynthesis, and have nested hierarchies of organisation. But the properties that make them similar fail to explain why each species is different.

Then there is a freedom and individuality in all particular things. A field of potatoes contains tens of thousands of genetically identical plants; cultivated potatoes are clones. Yet despite the fact they are in the same field, planted at the same time and experiencing the same weather, each plant is different from its neighbours; and each leaf on each plant is different in detail from every other leaf. Even the right and the left side of the same leaf have different patterns of veins and slightly different shapes.

The more the sciences generalise, the less they explain particulars, and vice versa. The sciences need to include both general principles and many specialised fields of study because the systems they investigate are so diverse, from quarks to galaxies, salt crystals to swallows' nests, and lichens to languages.

Scientific authority

One problem with the authority of science is that dissent and debate are dangerous. The need to preserve authority means that disagreements are usually kept behind the scenes. Scientists are reluctant to admit in public that their supposed objectivity can be compromised. Even Thomas Kuhn's theory of scientific revolutions as paradigm shifts preserved the image of established authority. In a scientific revolution, a new consensus reality replaces an old one. Ideas that were at first revolutionary become the new orthodoxy, like continental drift in geology, or quantum theory in physics. These are not like those rare political revolutions in which an autocratic system is overthrown and replaced by democracy. They are more like revolutions in which one dictatorship is replaced by another.

In almost every other sphere of human life, there is not one but many points of view. There are many languages, cultures, nations, philosophies, religions, sects, political parties, businesses and lifestyles. Only in the realm of science can we still find the old ethos of monopoly, universality and absolute authority that used to be claimed by the Roman Catholic Church. Catholic means 'universal'. At the Reformation, starting in 1517, the Roman Church lost its monopoly; now many other churches and ideologies coexist with it, including atheism. But there is still only one universal science.

In the seventeenth and eighteenth centuries, when Western Europe was divided by conflicts between Roman Catholics and Protestants, the ideals of science and reason shone out as a path to truth that rose above sectarian religious disputes. The Enlightenment grew out of this attitude of respect for the sciences and the power of human reason, accompanied by an attitude of condescension towards orthodox religion. As John Brooke wrote,

Science was respected not simply for its results, but as a way of thinking. It offered the prospect of enlightenment through the correction of past error, and especially through its power to override superstition . . . [But] the motivation of those who pitted science against religion often had little to do with gaining intellectual freedom for the study of nature. It was often not the natural philosophers [scientists] themselves, but thinkers with a social or political grievance, who transformed the sciences into a secularising force as they inveighed against clerical power.[14]

Scientists claimed to obtain absolute truth by viewing the world as objective observers.[15] In the black-and-white version of scientism, science is set apart from all other human activities. Science alone is capable of yielding unassailable facts.[16] In this idealised picture, scientists are exempt from the failings of the rest of humanity. They have a direct access to the truth. They are uniquely objective. The myths of disembodied knowledge and the allegory of the cave reinforce this image, and the prestige of the scientific priesthood adds the seal of authority.

This authoritarian mentality is most obvious in relation to psychic phenomena and alternative medicine (see Chapters 9 and 10). These are treated as heresies, rather than as valid areas for rational enquiry. Self-appointed inquisitions, like the Committee for Skeptical Inquiry, try to ensure that the subjects are not taken seriously in the respectable media, deprived of funding and excluded from university syllabuses. The belief that mechanistic medicine is the only kind that really works has far-ranging political consequences. There are many medical systems, including osteopathy, acupuncture, naturopathy and homeopathy, but only one kind, mechanistic medicine, is labelled 'scientific' and accorded a state-sponsored monopoly of power, scientific authority and financial support.

Science as we know it is based on an ideal of objective truth, allowing only one triumphant theory at a time. That is why scientists use phrases like 'knocking the final nail in the coffin of vitalism' (see page 13) or 'the final nail in the coffin of the steady state theory' (see page 67), gloating over the extermination of heresies. Much of the hypocrisy of science comes from assuming the mantle of absolute truth, which is a relic of the ethos of absolute religious and political power when mechanistic science was born. Of course, there are disagreements among scientists, and the sciences are continually changing and developing. But a monopoly of truth remains the ideal. Dissenting voices are heretical. Fair public debates are alien to the culture of the sciences.

In the Enlightenment ideal, science was a path to knowledge that would transform humanity for the better. Science and reason were in the vanguard. These were, and still are, wonderful ideals, and they have inspired scientists for generations. They inspire me. I am all in favour of science and reason if they are scientific and reasonable. But I am against granting scientists and the materialist worldview an exemption from critical thinking and sceptical investigation. We need an enlightenment of the Enlightenment.[17]

Scientific debates and dialogues

An important ingredient in the process of reform would be to introduce debates into scientific institutions. This may seem simple and obvious, but such debates are currently very rare. Debates are not yet part of the culture of science.

One potential debate that underlies much of this book is the question of whether the phenomena of life and mind can be reduced to physics. Many biologists believe they can. But many physicists are more doubtful. A debate on the subject 'Can the phenomena of life and mind be explained in terms of physics?' could happen on almost every university campus.

Another illuminating subject for debate would be the objectivity of the sciences. Universities and scientific institutes contain many people who put their faith in science and reason as a uniquely objective way of knowing. Many share Ricky Gervais's belief that 'Science is humble. It knows what it knows and it knows what it doesn't know. It bases its conclusions and beliefs on hard evidence.'[18] Many universities also contain historians, sociologists and philosophers of science who study how the sciences work in practice. They could debate how far the ideal of scientific objectivity corresponds to the practices of the sciences.

Then there are the ten fundamental dogmas of materialism discussed in Chapters 1 to 10 of this book. Each of them would make a good topic for debate, and I have suggested several further questions at the ends of all these chapters, most of which could provide topics for more specialised debates or dialogues.

If scientific debates became a normal feature of public life, university life and scientific conferences, the culture of science would change. Open questions would become normal, instead of one side being right and the other heretical. In democratic politics we are used to an ongoing pluralism, and no single party has a monopoly of public support. There are at least two sides to political arguments. In a democracy, the party in power cannot wipe out opposing views without becoming totalitarian and destroying the very principle of democracy.

But debates have their limitations, the main one being that one side wins the vote and the other loses. Likewise in courts of law, both sides argue their case, but the verdict goes one way or the other, yes or no. This system is invaluable when practical verdicts are needed. A judge and jury have to decide whether to convict or release someone accused of a crime. A parliament or a congress has to decide what laws to enact. There has to be one clear law or another, not a morass of legal ambiguity. Everyone has to drive on the right (as in the United States, France and Australia) or on the left (as in Britain, India and Japan). The decision may be arbitrary, but it has to be left *or* right, not left *and* right.

Some decisions in science have a similar practical necessity: which areas of research to fund, who gets a grant, whether to accept or reject a peer-reviewed paper for publication in a journal. The decisions are usually made in private, but there is often some kind of debate among the people who make the decisions.

All these practical debates, whether public or private, need to come to an agreed decision. But most situations are more ambiguous. At the frontiers of scientific research, when answers are not yet known, there is an inevitable uncertainty. Physicists do not agree whether one particular ten-dimensional string theory is correct, as opposed to other string theories and eleven-dimensional M-theories. Several different theories coexist, all with their advocates. In exploratory or uncertain areas, the most productive approach is not through debate but dialogue. A dialogue is an exchange of ideas or opinions, a joint exploration. It is not necessary for one side to win. Of course dialogues or conversations happen all the time in every walk of life, including among scientists, but if public dialogues became a regular aspect of scientific life they would encourage a culture of openness, even more than formal debates.

In my experience, the most productive dialogues are between two or three people.[19] So-called panel discussions, a standard feature of scientific conferences, with five to ten participants, rarely achieve anything. By the time each participant has made an opening statement, there is usually no time left for discussion, and with so many participants a clear focus is often impossible. Two or three people can go further faster.

Public participation in science funding

Science has always been élitist and undemocratic, whether in monarchies, Communist states or liberal democracies. But it is currently becoming more hierarchical, not less so. In the nineteenth century, Charles Darwin was one of many independent researchers who, not reliant on grants, did provocatively original work. That kind of freedom and independence is rare today. Science-funding committees determine what can happen in research. The power in those committees is concentrated in the hands of politically adept older scientists, government officials and representatives of big business.

In 2000, a government-sponsored survey in Britain on public attitudes to science revealed most people believed that 'Science is driven by business – at the end of the day it's all about money.' More than three-quarters of those surveyed thought, 'It is

important to have some scientists who are not linked to business.'
More than two-thirds said, 'Scientists should listen more to what
ordinary people think.' Worried about this public alienation, the
British government tried to engage the wider public in 'a dialogue
between science, policy-makers and the public'.[20] In official circles,
the fashion shifted from the previous policy of the public understanding of science to an 'engagement' model of science and
society. The public-understanding policy was based on a 'deficit'
model, which saw simple factual education as the key. Scientists
should tell the public the truth, and they would accept it gratefully.
The trouble was that this policy did not work. The British public
was told that mad cow disease was no threat to humans. Then it
was. Then they were told that genetically modified (GM) crops
were good for them, and many did not believe it. Throughout
Europe there was a consumer revolt against GM foods, and proponents of the public understanding of science were powerless to
prevent it.

'Public engagement' with science was supposed to be the
answer. But this change in rhetoric made little difference in practice, and the funding of science carried on as before. So did public
distrust. And although there were several well-organised public-
engagement exercises in the 2000s, policy-makers usually ignored
them.[21]

Some of the few examples of effective engagement are in medicine, where patient activist groups, like AIDS activists, have
already had a major impact on research and treatment.[22] There are
many kinds of patient group. Some are primarily mutual help
organisations, while others are highly politicised. Sociologists
who study these groups have suggested that they exemplify the
emergence of 'scientific citizenship'.[23] However, some patient
groups are funded by the pharmaceutical companies, who stand
to gain from campaigns for health providers to pay for expensive
drugs. But despite this exploitation of some patient groups, many
of these organisations demonstrate that lay people are well able to
participate in technical discussions.

Medical research charities, like Cancer Research UK, the

Meningitis Research Foundation and the Stroke Association, have a direct influence on research by funding it. In the UK there are 130 such charities,[24] and collectively they contribute about one-third of all public expenditure on medical and health research. Some are governed by boards or committees mainly composed of lay people.

The interests of patient activist groups and medical charities are confined to particular diseases and disabilities. For people without such an intense focus, there is at present little possibility of engaging with scientific research. I suggest an experiment that would make more widespread public engagement a reality. Spend one per cent of the science budget on research that actually interests people outside the scientific and medical professions. At present, money is allocated according to agendas set by committees of establishment scientists, corporate executives and government bureaucrats. In the UK, these official funding bodies include the Medical Research Council, the Biotechnology and Biological Sciences Research Council and the Engineering and Physical Sciences Research Council. The UK government's scientific research budget is about £4.6 billion per year,[25] so the one per cent fund would contain about £46 million a year.

What questions capable of being answered by scientific research are of public interest? The simplest way to find out would be to ask for suggestions. They could come from membership organisations like the National Trust, the British Beekeepers' Association, the National Society of Allotment and Leisure Gardeners, Oxfam, the Consumers' Association, the Women's Institute, as well as local authorities and trade unions. Potential subjects for research would be discussed in these organisations' newsletters, in specialist magazines, newspapers and in online forums. Their research suggestions would be submitted to the body that administers the one per cent fund, which could be called the Open Research Centre.

The Open Research Centre would be independent of the science establishment, and governed by a board representing a wide range of interests, including non-governmental organisations and

voluntary associations. Like some of the medical research charities, most of its members would be non-scientists. Based on the
suggestions it received, it would publish a list of the research areas
in which grants were available, and would invite proposals that
would be evaluated by experts in the usual way. It would not fund
research already covered by the regular science budget.

This new venture, open to democratic input and public participation, would involve no additional expenditure, but would have
a big effect on people's involvement in science and innovation.[26] I
expect it would make the sciences more attractive to young people,
stimulate public interest in scientific thinking, and help break
down the depressing alienation many people feel from the sciences.
It would enable scientists themselves to think more freely. And it
would be more fun.

In addition, there could be other engaging methods of funding
scientific projects. One possibility would be a reality TV show in
which proposals for research of widespread public interest are
submitted to a panel, rather like the BBC TV show *Dragons' Den*,
in which entrepreneurs pitch for investments from a panel of business people. The panel, including both scientists and non-scientists,
would have real money to give out as grants – say £1 million a year,
taken from the one per cent fund.

The greater the diversity of funding sources, the greater the
freedom of the sciences. Fortunately, there is already a range of
non-governmental sources of funding including businesses and
charitable foundations, and some of them already fund areas of
research that are taboo for official funding agencies. Foundations
have more freedom to adapt to new circumstances than government funding agencies, and may be in the best position to facilitate
the opening up of new lines of research.

Learning from other cultures

The sciences as we know them are weakest when they are dealing
with, or trying to avoid, the subjective aspects of reality. Our own
experience of qualities like the smell of a rose or the sound of a

band has been stripped away, leaving only odourless molecular structures and the physics of vibrations. The sciences have tried to confine themselves to I-it relationships, a third-person view of the world. They have done their best to leave out I-you relationships, second-person experiences, as well as first-person experiences, our personal experiences. Our inner life, including our dreams, hopes, loves, hates, pains, excitements, intentions, joys and sorrows, is reduced to charts of readings from electrodes, as in electroencephalograms (EEGs), or changes in the levels of the chemicals at nerve endings, or 2-D brain scans on computer screens. By these means a mind becomes an 'it', an object.

But instead of trying to reduce minds to objects, what if all self-organising systems are subjects? As discussed in Chapter 4, some philosophers propose that materialism implies panpsychism, meaning that self-organising systems like atoms, molecules, crystals, plants and animals have points of view, or inner lives, or subjective experience. Most people who keep companion animals take for granted that their dog or their cat or their parrot or their horse has subjective experiences, like emotions, desires and fears. But what about snakes? Or oysters? Or plants? We can try to imagine their inner lives, but it is difficult to do so. Yet in traditional hunter-gatherer societies all around the world, specialists in communication with non-human organisms form connections with a wide range of animals and plants. Shamans link themselves to animals and plants through their minds or spirits, and find out useful information by doing so. They are said to know where animals are to be found, and they help hunters. They know which plants are useful in healing, or as mind-altering brews.

For centuries, among scientists and educated people in the West, shamanic knowledge has been dismissed as primitive, animistic or superstitious. Anthropologists have studied the social roles of shamans, but most of them have assumed that if shamans have any valid knowledge of the natural world, it has not been gained subjectively but rather by 'normal' sense-based means, or by trial and error. They think that if shamans have discovered herbs that work, or visionary brews like *ayahuasca*, traditionally used in parts of the Amazon region, they have done so by trying out various plants at random. But shamans themselves say that this knowledge has come from 'the plant teachers'.[27]

What if shamans really do have ways of learning about plants and animals that are completely unknown to scientists? What if they have explored the natural world for many generations, discovering ways of communicating with the world around them that depend on subjective rather than objective methods? The Brazilian anthropologist Viveiros de Castro summarised the difference:

> Objectification is the name of our game . . . The form of the other is *the thing*. Amerindian shamanism is guided by the opposite ideal. To know is to personify, to take on the point of view of that which must be known. Shamanic knowledge aims at something that is a someone – another subject. The form of the other is *the person*. What I am defining here is what anthropologists of yore used to call animism, an attitude that is far more than an idle metaphysical tenet, for the attribution of soul to animals and other so-called natural beings entails a specific way of dealing with them.[28]

For most of human history, people have lived as hunter-gatherers, and have only survived because they knew how to hunt and had a deep understanding of the animals they hunted. They only survived because they knew which plants were edible, and where and when to find them. Their knowledge worked. We still benefit from their discoveries. About 70 per cent of our drugs are ultimately derived from plants (see Chapter 10), and much of the knowledge of these plants' medicinal properties was traditional, discovered long ago in pre-scientific cultures.

For much of the twentieth century, scientific psychologists tried to learn about minds objectively, from outside, by studying measurable behaviour and quantifiable responses. In prototypical behaviourist experiments, rats in cages learned to press levers to obtain rewards in the form of food pellets or to avoid punishments like electric shocks. In more recent research, the emphasis has mainly been on the study of brains and of computer models of brain activity. In the mystical traditions of both East and West, people explored the nature of minds through long periods of meditation, discovering how their mental processes work from within. By contrast, academic psychologists and cognitive scientists usually carry out their studies with paid subjects, generally undergraduate students, who have no professional training in observing or reporting mental processes. As the Buddhist scholar B. Alan Wallace put it:

> By leaving introspection in the hands of amateurs, scientists guarantee that the direct observation of the mind remains at the level of folk psychology . . . Cognitive scientists have taken on the challenge of understanding mental processes, but unlike all other natural scientists, they receive no professional training in observing the realities that constitute their field of enquiry.[29]

Today there are many teachers of meditation, mainly rooted in the Hindu and Buddhist traditions, and some scientists have begun to explore their own minds for themselves.[30]

Scientific investigations of the interactions of minds and bodies are as backward as the investigation of minds from within. In medicine, there is a growing recognition of the effects of belief on healing, as shown in the placebo response, and studies using biofeedback show that people can learn to gain conscious control over their blood flow and their fingers and other aspects of their physiology that are normally regulated unconsciously (see Chapter 10). But these achievements are elementary compared with the feats of Indian yogis, who demonstrate a remarkable voluntary influence on their digestive and circulatory systems. One of the means by which they acquire these abilities is through the control of breathing. Breathing is controlled by both the voluntary and involuntary nervous systems, and yogic breathing exercises may provide a bridge between them.[31]

In China, the *chi gung* or *qigong* tradition likewise places a strong emphasis on breathing practices, and has many applications in traditional Chinese medicine and in the martial arts. Both *prana* in the Indian tradition and *chi* in the Chinese are translated into English as 'energy' but they differ from the concept of energy in mechanistic physiology. There are serious problems with the standard scientific dogma of energy conservation in living organisms (see Chapter 2) and a re-examination of human energy balances is long overdue. This is one area in which it might be possible to bring together these different traditions in a new, integrated understanding.

In many parts of Africa and the India subcontinent, women carry heavy loads on their heads, and can do so over great distances. Studies of women in East Africa have shown that they can carry up to 20 per cent of their body weight 'for free', without any extra expenditure of energy compared with just walking. They can also carry up to 70 per cent of their body weight using 50 per cent less energy than an American army recruit with a backpack. This skill is not simply a matter of putting the load on the head, but involves a special kind of gait.[32] But is a special gait enough to explain this remarkable efficiency?

Their abilities also raise a practical question. Why are teenagers not taught this skill in physical-education classes all over the world? The ability to carry loads efficiently is useful. At some stage in their lives, modern people may need to carry loads over rougher terrain than they encounter in airports, when wheeled suitcases will not work. The main reason for ignoring this skill is social status. Women who carry loads on their heads are of low status, and live in developing countries.

Arrogance and snobbery make most modern, scientifically educated people feel superior to all pre-scientific cultures, including their own. In the late nineteenth century, these attitudes were given a scientific justification in terms of evolution and social progress. Anthropologists, like James Frazer (1854–1941), thought that human beliefs progressed through three stages: animism, religion and science. Primitive societies were animistic and child-like, pervaded by magical thinking. Religions like Christianity represented a higher stage of evolution, but still included many primitive elements. But both animism and religion were super-seded by science, the ultimate level of human understanding.

In this context, why would modern people want to learn to carry loads on their heads, like uneducated African women? Or why would they have anything to learn from pre-scientific tradi-tions like yoga and *chi gung*? And what have shamans to offer but mumbo-jumbo?

New dialogues with religions

As the sciences free themselves from the constrictions of material-ism, many new possibilities arise. And many of them raise new possibilities for dialogues with religious traditions.[33] Here are a few examples.

Statistical research has shown that people who attend religious services regularly tend to live longer, have better health and are less prone to depression than those who do not. Also, the practices of prayer and meditation often have beneficial effects on health and longevity (see Chapter 10). How do these practices work? Are the effects purely psychological or sociological? Or does the connection with a larger spiritual reality confer a greater capacity to heal and an enhancement of wellbeing?

If organisms at all levels of complexity are in some sense alive with their own purposes, this implies that the earth, the solar system, our galaxy and, indeed, all the stars have lives and purposes of their own. And so may the entire universe (see Chapter 1). The cosmic evolutionary process may have inherent purposes or ends, and the cosmos may have a mind or consciousness. Since the universe itself is evolving and developing, the mind or conscious-ness of the universe must be evolving and developing too. Is this cosmic mind the same as God? Perhaps only if God is conceived of in a pantheistic spirit as the soul or mind of the universe, or of nature. In the Christian tradition, the world soul is not identical with God. For example, the early Christian theologian Origen (*c.* 184–253) thought of the world soul as the Logos, endlessly creative, which gave rise to the world and the processes of development within it. The Logos was an aspect of God, not the whole of God, whose being transcended the universe.[34] If instead of one universe there are many, then the divine being would include and transcend them all.

The universe is evolving and is the arena of continuing creativity. Creativity is not confined to the origin of the universe, as in deism (see Chapter 1), but is an ongoing part of the evolutionary process, expressed in all realms of nature, including human societies, cultures and minds. Although the creativity expressed in all these realms may have an ultimately divine source, there is no need to think of God as an external designing mind. In the Judaeo-Christian tradition, God imbued the natural world with creativity too, as in the first chapter in the Book of Genesis, where he called forth life from the earth and the seas (Genesis 1: 11, 20, 24) – a very different image from the engineering God of a mechanistic universe. And in a creative, evolving universe there is no reason why the appearance of matter and energy should be confined to the very first instant, as in the standard Big Bang theory. Indeed, some cosmologists propose that the continued expansion of the universe is driven by the ongoing creation of 'dark energy' from the universal gravitational field or from the 'quintessence field' (see Chapter 2).

If the laws of nature are more like habits, and there is an inherent memory within the natural world (see Chapter 3), how does this relate to the principle of *karma* in Hinduism and Buddhism, a chain of cause and effect that implies a kind of memory in nature? In some schools of thought, as in the *Lankavatra Sutra* of Mahayana Buddhism, there is a cosmic or universal memory.[35] Likewise, if biological inheritance largely depends on morphic resonance and a collective memory within each species (see Chapter 6), how does this relate to doctrines of reincarnation or rebirth?

If minds are not stored as material traces in brains, but depend on a process of resonance, then memories themselves may not be extinguished at death, although the body through which they are normally retrieved decays. Is there some other way in which these memories can continue to act? Can some non-bodily form of consciousness survive the death of the body and still gain access to an individual's memories, conscious or unconscious, as all religions suppose?

If minds are not confined to brains, how do these human minds relate to the minds of higher-level systems of organisation, like the solar system, the galaxy, the universe and the mind of God? Are mystical experiences just what they seem to be: connections between human minds and larger, more inclusive forms of consciousness?

If human minds, individually and collectively, make contact with higher-level minds, including the ultimate consciousness of God, to what extent can they influence the evolutionary process, or be influenced by the divine will? In an evolutionary, living universe, are humans merely part of an unfolding process on one isolated planet, or does human consciousness play a larger role in cosmic evolution, in some way connected to minds in other parts of the universe?

All religious traditions grew up in a pre-scientific era. The sciences have revealed far more of the natural world than anyone could have imagined in the past. For example, only in the nineteenth century were the great sweep of biological evolution and the aeons of geological times recognised, and only in the twentieth century were galaxies outside our own discovered, along with the vast expanse of time from the Big Bang to the present. The sciences evolve, and so do religions. No religion is the same today as it was at the time of its founder. Instead of the bitter conflicts and mutual distrust caused by the materialist worldview, we are entering an era in which sciences and religions may enrich each other through shared explorations.

Open questions

As the taboos of materialism lose their power, new scientific questions can be asked and, hopefully, answered.

Throughout this book, I have suggested a range of new possibilities for research: for example, the use of comparative effective research on conventional and 'alternative' cures for conditions such as lower back pain, migraines and cold sores (see Chapter 10); experiments on experiments to find out how significantly experimenters' expectations influence their results in the 'hard' sciences (see Chapter 11); an analysis of existing data to find out if the Universal Gravitational Constant varies (see Chapter 3); a mass-participation investigation to find out if earthquakes and tsunamis can be predicted on the basis of animal premonitions (see Chapter 9); and a prize challenge to find out if any alternative energy technologies or 'over unity' devices actually work (see Chapter 2).

Existing lines of scientific research will, of course, continue. Nothing changes very fast when big institutions, vast amounts of money and large numbers of jobs are involved: there are now more than seven million scientific researchers worldwide, producing 1.58 million publications a year.[36] What I am suggesting is that a small fraction of these resources is devoted to exploring new questions. New discoveries are more likely to happen if we venture off the well-trodden paths of conventional research, and if we open up questions that have been suppressed by dogmas and taboos.

The delusion that science has already answered the fundamental questions chokes off the spirit of enquiry. The illusion that scientists are superior to the rest of humanity means that they have little to learn from anyone else. They need other people's financial support, but they do not need to listen to anyone less scientifically educated than themselves. In return for their privileged position, scientists will deliver knowledge and power over nature, transforming humanity and the earth.

The materialist agenda was once liberating but is now depressing. Those who believe in it are alienated from their own experience; they are cut off from all religious traditions; and they are prone to suffer from a sense of disconnection and isolation. Meanwhile, the power unleashed by scientific knowledge is causing the mass extinction of other species, and endangering our own.

The realisation that the sciences do not know the fundamental answers leads to humility rather than arrogance, and openness rather than dogmatism.

Much remains to be discovered and rediscovered, including wisdom.

Notes

Preface

1. This work is reviewed in Sheldrake (1973).
2. Rubery and Sheldrake (1974).
3. Sheldrake and Moir (1970).
4. Sheldrake (1974).
5. Sheldrake (1984).
6. E.g., Sheldrake (1987).

Introduction: The Ten Dogmas of Modern Science

1. In Popper and Eccles (1977).
2. E.g., D'Espagnat (1976).
3. Hawking and Mlodinow (2010), p. 117.
4. Ibid., pp. 118–19.
5. Smolin (2006).
6. Carr (ed.) (2007); Greene (2011).
7. Ellis (2011).
8. Collins, in Carr (ed.) (2007), pp. 459–80.

Prologue: Science, Religion and Power

1. Ibid., p. 50.
2. Bacon (1951), pp. 290–91.
3. Ibid., p. 298.

4. Fara (2009), p. 132.
5. Kealey (1996).
6. Dubos (1960), p. 146.
7. Kealey (1996).
8. National Science Board (2010), Chapter 4.
9. Sarton (1955), p. 12.
10. Laplace (1819), p. 4.
11. Ibid.
12. Chivers (2010).
13. Munowitz (2005), Chapter 7.
14. Chivers (2010).
15. Gould (1989).
16. Gleik (1988).
17. Malhotra *et al.* (2001).
18. Quoted in Horgan (1997b).
19. Horgan (1997b), p. 6.
20. Westfall (1980).
21. Burtt (1932).
22. Gould (1999).
23. Quoted in Burtt (1932), p. 9.
24. Kekreja (2009).
25. Wikipedia: *The God Delusion*, accessed 16 June 2011: http://en.wikipedia.org/wiki/The_God_Delusion
26. Gray (2007), pp. 266–7.
27. Gray (2002), p. xiii.
28. Kuhn (1970).
29. Latour (1987), pp. 184–5.
30. Gervais (2010).

1: Is Nature Mechanical?

1. Quoted in Brooke (1991), p. 120.
2. Ibid., p. 119.
3. Burtt (1932), p. 45.
4. Ibid., p. 120.
5. Quoted in Collins (1965), p. 81.
6. Burtt (1932), p. 73.
7. Wallace, trans. (1911), p. 80.
8. Brooke (1991), pp. 128–9.
9. Descartes (1985), Vol. 1, p. 317.
10. Ibid., p. 139.
11. Ibid., p. 131.
12. Ibid., p. 141.
13. Dennett (1991), p. 43.
14. Kretzman and Stump (1993).
15. Gilson (1984).
16. Gilbert (1600).
17. Sheldrake (1990), Chapter 4.
18. Lightman (2007), p. 188.
19. Burtt (1932).
20. Descartes (1985), Vol. 1, p. 101.
21. Kahn (1949).
22. E.g., Wiseman (2011), pp. 74, 77, 81, 93, 108, 128, 169.
23. Grayling (2011) made these remarks while paraphrasing arguments by Michael Smermer in a book called *The Believing Brain*, endorsing them as likely to be 'the right view'.
24. E.g., Shermer (2011).
25. Brooke (1991), p. 134.
26. Ibid., p. 146.
27. Paley (1802).
28. Quoted in Lightman (2007), p. 45.
29. Dembski (1998).
30. Brown *et al.* (1968), p. 11.
31. Schelling (1988).
32. Richard, in Cunningham and Jardine (eds) (1990), p. 131.
33. Wroe (2007).
34. Bowler (1984), pp. 76–84.
35. Darwin (1794–6).
36. Lamarck (1914), p. 122.
37. Ibid., p. 36.
38. Bowler (1984), p. 134.
39. Darwin (1875), pp. 7–8.
40. Darwin (1859), Chapter 3.
41. Particularly in Darwin (1875).
42. Monod (1972).
43. Partridge (1961), pp. 386–7.
44. Quoted in Driesch (1914), p. 119.
45. Huxley (1867).
46. Dawkins (1976), p. 22.
47. Ibid., p. 21.
48. Ibid., Preface.
49. Dawkins (1982), p. 15.
50. Smuts (1926).
51. Ibid., Chapter 12.
52. Ibid., p. 97.
53. Whitehead (1925), Chapter 6.
54. Koestler (1967), p. 385.
55. Banathy (1997), http://www.newciv.org/ISSS_Primer/asem04bb.html
56. Mitchell (2009).
57. Filippini and Gramaccioli (1989).
58. Hume (2008), Part VII.
59. Thomson (1852).
60. Singh (2004).
61. Long (1983).

2: Is the Total Amount of Matter and Energy Always the Same?

1. Burnet (1930).
2. Dijksterhuis (1961), p. 9.
3. Tarnas (1991).
4. Ibid., p. 437.
5. Newton (1730, reprinted 1952), Query 31, p. 400.
6. Popper and Eccles (1977), p. 5.
7. Ibid., p. 7.
8. Davies (1984), p. 5.
9. Munowitz (2005).
10. Coopersmith (2010), p. 23.
11. Ibid., p. 255.
12. Ibid., p. 265.
13. Kuhn (1959).
14. For an excellent history of concepts of energy, see Coopersmith (2010).
15. Harman (1982), p. 58.
16. Feynman (1964).
17. Sheldrake, McKenna and Abraham (2005).
18. Quoted in Singh (2004), p. 360.
19. William Bonner, quoted by Singh (2004), p. 361.
20. Singh (2004).
21. Quoted by Singh (2004), p. 418.
22. Singh (2004).
23. Ibid., p. 133.
24. E.g., Bekenstein (2004).
25. Singh (2004), p. 139.
26. Belokov and Hooper (2010).
27. Coopersmith (2010), p. 20.
28. Ibid., p. 292.
29. Thomson (1852).
30. Quoted in Burtt (1932), p. 9.
31. Davies (2006), Chapter 6.
32. Ostriker and Steinhardt (2001).
33. Sobel (1998).
34. http://www.xprize.org/
35. Coopersmith (2010), pp. 270–79.
36. Quoted by Coopersmith (2010), p. 329.
37. Frankenfield (2010).
38. Webb (1991).
39. Ibid.
40. Webb (1980).
41. Webb (1991).
42. Webb (1980).
43. Frankenfield (2010), p. 947.
44. Ibid., p. 1300.
45. Webb (1991).
46. Dasgupta (2010).
47. Thurston (1952).
48. Ibid., p. 377.
49. Ibid., p. 366.
50. Ibid., p. 384.

3: Are the Laws of Nature Fixed?

1. Tarnas (1991), p. 46.
2. Plato, *The Republic*, Book 7.
3. Tarnas (1991), p. 47.
4. Burtt (1932), p. 64.
5. Quoted in Pagels (1983), p. 336.
6. In Wilber (ed.) (1984), p. 185.
7. Ibid., p. 137.
8. Ibid., p. 51.

9. For data, see Sheldrake (1994), Chapter 6.

10. Mohr and Taylor (2001).

11. Schwarz *et al.* (1998).

12. References for measurements at different dates: 1973: Cohen and Taylor (1973); 1986: Holding *et al.* (1986); 1988: Cohen and Taylor (1988); 1995: Kiernan (1995); 1998: Schwarz *et al.* (1998); 2000: Grundlach and Merkowitz (2000); 2010: Reich (2010).

13. Schwarz *et al.* (1998).

14. Stephenson (1967).

15. For a discussion, see Sheldrake (1994), Chapter 6.

16. Brooks (2009), Chapter 3.

17. Adam (2002).

18. Brooks (2010).

19. Barrow and Webb (2005).

20. Birge (1929), p. 68.

21. For data and references, see Sheldrake (1994), Chapter 6.

22. De Bray (1934).

23. Petley (1985), p. 294.

24. Davies (2006).

25. Ibid.

26. Hawking and Mlodinow (2010), p. 118.

27. Tegmark (2007), p. 118.

28. Rees (1997), p. 3.

29. Ibid., p. 262.

30. Woit (2007).

31. Smolin (2006).

32. Ibid.

33. Bojowald (2008).

34. Smolin (2010).

35. Robertson *et al.* (2010).

36. Quoted in Potters (1967), p. 190.

37. Ibid.

38. Nietzsche (1911).

39. In Murphy and Ballou (1961).

40. Whitehead (1954), p. 363.

41. Sheldrake (1981, new edition 2009).

42. In Sheldrake (2009), Appendix B.

43. Bohm (1980), p. 177.

44. *Cf.* Laszlo (2007).

45. *Cf.* Carr (2008).

46. Woodard and McCrone (1975).

47. Ibid.

48. Holden and Singer (1961), pp. 80–81.

49. Ibid., p. 81.

50. Woodard and McCrone (1975).

51. Goho (2004).

52. Bernstein (2002), p. 90.

53. Quoted in Woodard and McCrone (1975).

54. Danckwerts (1982).

55. Sheldrake (2009).

56. Bergson (1946), p. 101.

57. Ibid., pp. 104–5.

58. Bergson (1911), p. 110.

4: Is Matter Unconscious?

1. Dennett (1991), p. 37.

2. Crick (1994), p. 3.

3. Griffin (1998).

4. Huxley (1893), p. 240.

5. Ibid., p. 244.

6. The most ingenious evolutionary arguments for the

emergence of illusory consciousness are by Humphrey (2011).

7. Searle (1992), p. 30.
8. Strawson (2006), p. 5.
9. Crick (1994), pp. 262–3.
10. Strawson (2006).
11. Ibid.
12. Ibid., p. 27.
13. Searle (1997), pp. 43–50.
14. Spinoza (2004), Part III, propositions 6–7.
15. Hampshire (1951), p. 127.
16. Skrbina (2003).
17. Ibid., p. 20.
18. Ibid., p. 21.
19. Ibid., p. 21.
20. Ibid., p. 22.
21. Ibid., p. 25.
22. Ibid., p. 27.
23. Ibid., p. 28.
24. Ibid., p. 31.
25. Ibid., p. 32.
26. Ibid., p. 33.
27. Dennett (1991), pp. 173–4.
28. Griffin (1998), p. 49 note.
29. Ibid.
30. Ibid., p. 113.
31. De Quincey (2008).
32. Ibid.
33. Ibid., p. 99.
34. Libet et al. (1979), p. 202.
35. Libet (1999).
36. Wegner (2002).
37. Libet (2006).
38. Libet (2003), p. 27.
39. Feynman (1962).
40. Quoted by Dossey (1991), p. 12.
41. Dyson (1979), p. 249.

5: Is Nature Purposeless?

1. Dawkins (1976).
2. Haemmerling (1963).
3. Goodwin (1994), Chapter 4.
4. Hinde (1982).
5. Smith (1978).
6. Thom (1975, 1983). Note that Thom spells chreode 'chreod' without the final 'e'. Chreode with an 'e' is Waddington's original spelling.
7. Thom (1975).
8. Cramer (1986).
9. Aharonov et al. (2010).
10. Anfinsen and Scheraga (1975).
11. In an ongoing series of workshops on predicting protein structures held under the aegis of the Lawrence Livermore National Laboratory in California, teams from all over the world try to predict the three-dimensional structure of proteins without knowing the answer. These evaluations are called the Critical Assessment of Techniques for Protein Structure Prediction (CASP). By far the most successful predictions are based on a detailed knowledge of similar proteins, known as comparative modelling. The CASP competitions used to include an ab initio category, implying that the predictions started from first principles, but for

CASP6 in 2004, the name of the category was changed: 'This name implies that there is no reliance on known structures in building models. In practice, most of the methods used for such targets do make extensive use of available structural information, both in devising scoring functions to distinguish between correct and incorrect predictions, and in choosing fragments to incorporate in the model. For this reason, the category was renamed as new folds.' www.prediction-center.org/casp6/doc/categories.html

12. For a review, see Nemethy and Scheraga (1977).
13. Anfinsen and Scheraga (1975).
14. *Cf.* Elsasser's (1975) 'principle of finite classes'.
15. Hawking (1988), p. 60.
16. Smolin (2006).
17. Thom (1975), pp. 113–14, 141.
18. Ibid., Chapter 9.
19. For a general introduction, see Capra (1996).
20. Thom (1983), p. 141.
21. Penrose (2010).
22. Bergson (1911), p. 262.
23. Cohn (1957).
24. Bacon (1951).
25. Midgley (2002), Chapter 7.
26. Satprem (2000).

6: Is All Biological Inheritance Material?

1. Cole (1930).
2. Needham (1959), p. 205.
3. Holder (1981).
4. Dawkins (1976).
5. Ibid., p. 23.
6. Ibid., p. 24.
7. Hodges (1983).
8. E.g., Carroll (2005), p. 106.
9. For accounts of the vitalist-mechanist controversies, see Nordenskiold (1928); Coleman (1977).
10. Venter (2007), p. 299.
11. Ibid., p. 300.
12. Ibid.
13. Quoted in *Nature* (2011).
14. Ibid.
15. Culotta (2005).
16. Manolio *et al.* (2009).
17. Khoury *et al.* (2010).
18. Green and Guyer (2011).
19. Latham (2011).
20. *Wall Street Journal,* 2 May 2004.
21. Pisano (2006), p. 184.
22. Ibid., p. 198.
23. Howe and Rhee (2008).
24. Carroll *et al.* (2001).
25. Gerhart and Kirschner (1997).
26. Wolpert (2009). The Edge Question Center, 2009 http://www.edge.org/q2009/q09_6.html#wolpert
27. http://www.sheldrake.org/D&C/controversies/genomewager.html

28. Wolpert and Sheldrake (2009). See also Schnabel (2009).
29. Darwin (1859; 1875).
30. Mayr (1982), p. 356.
31. Ibid., Chapter. 5.
32. Huxley (1959), p. 8.
33. Ibid., p. 489.
34. Medvedev (1969).
35. Anway et al. (2005).
36. Young (2008).
37. Petronis (2010).
38. Qiu (2006).
39. Galton (1875).
40. Quoted in Wright (1997), p. 17.
41. Ibid., p. 21.
42. Wright (1997), Chapter 2.
43. Iacono and McGue (2002).
44. Watson (1981).
45. Wright (1997), p. 42.
46. Dawkins (1976), p. 206.
47. E.g., Blackmore (1999).
48. Dawkins, in Blackmore (1999), p. ix.
49. Ibid.
50. E.g., Blackmore (1999), Dennett (2006).
51. Sheldrake (2011b).
52. This conversation took place at Ashton Wold, the house of Dame Miriam Rothschild, in the summer of 1995 or 1996.
53. Conniff (2006).
54. Dawkins (2006, p. 215) wrote that he was 'mortified' to learn that The Selfish Gene had inspired Jeffrey Skilling and other Enron executives, and thought that they had misunderstood his message.

7: Are Memories Stored as Material Traces?

1. Rose (1986), p. 40.
2. Plotinus (1956), Ennead 4, Tractate 6.
3. Inge (1929), Vol. 1, pp. 226–8.
4. Bursen (1978).
5. Crick (1984).
6. Boakes (1984).
7. Lashley (1929), p. 14.
8. Lashley (1950), p. 479.
9. Pribram (1971); Wilber, ed. (1982).
10. Boycott (1965), p. 48.
11. Rose and Harding (1984); Rose and Csillag (1985); Horn (1986); Rose (1986). In similar experiments with chicks, detailed studies have showed that there are changes in the number of vesicles in the synapses following learning (Rose, 1986).
12. Cipolla-Neto et al. (1982).
13. Kandel (2003).
14. Lu et al. (2009).
15. Lewin (1980).
16. Fröhlich and McCormick (2010).
17. Blackiston et al. (2008).
18. Lashley (1950), p. 472.
19. Hunter (1964).
20. E.g., Squire (1986). For vivid descriptions of some clinical cases see Sacks (1985).
21. Luria (1970; 1973); Gardner (1974).
22. Penfield and Roberts (1959).

23. Quoted in Wolf (1984), p. 175.
24. Pribram (1979).
25. Bohm (1980).
26. Bohm in Weber (1986), p. 26.
27. Bohm in Sheldrake (2009), p. 302.
28. Morphic resonance and the evidence for it is discussed in detail in my book *A New Science of Life* (new edition 2009). Its historical background and wider implications are explored in my book *The Presence of the Past: Morphic Resonance and the Habits of Nature* (new edition 2011).
29. Jennings (1906).
30. Wood (1982).
31. Wood (1988).
32. Klein and Kandel (1978).
33. Jennings (1906).
34. Watkins *et al.* (2010).
35. Rizzolatti *et al.* (1999).
36. Agnew *et al.* (2007).
37. Ibid., p. 211.
38. Yates (1969).
39. E.g., Lorayne (1950).
40. I discuss the details of these experiments in my book *A New Science of Life* (new edition 2009), with all relevant references to the original papers in scientific journals.
41. Flynn (2007).
42. Data from Horgan (1997a).
43. Flynn (2007), p. 176.
44. Summarized in Sheldrake (2009).

8: Are Minds Confined to Brains?

1. Piaget (1973), p. 280.
2. Wallace (2000), pp. 28–9.
3. Ibid., p. 49.
4. Crick (1994), p. 3.
5. Greenfield (2000), pp. 12–15.
6. The neurologist Wilder Penfield found that he could evoke vivid flashes of memory by stimulating the cerebral cortex of patients during brain operations, but although this stimulation could evoke memories, he did not think they were located in the part stimulated, and he also concluded that the memory 'is not in the cortex' (Penfield, 1975).
7. Duncan and Kennett (2001), p. 8.
8. Lindberg (1981).
9. Ibid., p. 202.
10. Kandel *et al.* (1995) p. 368.
11. Gray (2004), pp. 10, 25.
12. Lehar (2004).
13. Lehar (1999).
14. Winer *et al.* (2002).
15. Winer *et al.* (1996).
16. Winer and Cottrell (1996).
17. Ibid. (1996).
18. E.g., Bergson (1911); Burtt (1932).
19. James (1904), quoted in Velmans (2000).
20. Whitehead (1925), p. 54.
21. Velmans (2000), p. 109.
22. Ibid., pp. 113–14.
23. Gibson (1986).

24. Thompson *et al.* (1992).
25. Noë (2009), p. 183.
26. In Blackmore (2005), p. 164.
27. Bergson (1911), p. 7.
28. Ibid., pp. 37–8.
29. Sheldrake (2005b).
30. Braud *et al.* (1990); Sheldrake (1994); Cottrell *et al.* (1996).
31. Sheldrake (2003a).
32. Ibid.
33. Ibid.
34. Ibid.
35. Ibid.
36. Corbett (1986); Sheldrake (2003a).
37. Long (1919).
38. Cottrell *et al.* (1996).
39. Sheldrake (2003a).
40. Ibid.
41. Sheldrake (2005a).
42. The statistical significance was p= 10^{-376} (Sheldrake, 2005a).
43. Ibid.
44. Sheldrake (2003a).
45. In a meta-analysis of fifteen CCTV staring studies, most showed positive effects and the overall positive effect was statistically significant (Schmidt *et al.*, 2004).
46. Dyson (1979).
47. Ibid., p. 171.

9: Are Psychic Phenomena Illusory?

1. Quoted in Barrett (1904).
2. In Krippner and Friedman (eds) (2010).
3. Quoted in Auden (2009).
4. *New Penguin English Dictionary*, 1986.
5. For discussions, see Sheldrake (2003a, 2011) and Radin (1997, 2007).
6. Recordon *et al.* (1968).
7. As Peters pointed out, the only possible 'normal' explanation could be that the mother was somehow sending the boy some kind of secret or unconscious auditory code over the telephone, but there was no evidence that she could have been doing this. In any case, Peters provided the tape recordings to anyone who was interested so they could try to detect cues themselves. I listened to the tapes and there was no trace of any possible code, nor could a professional magician detect any possible kind of cheating.
8. Sheldrake (2003a).
9. Radin (1997).
10. Ibid.
11. Ibid.
12. Ullman, Krippner and Vaughan (1973).
13. The odds against chance were 75 million to one (Radin, 1997).
14. Carter, in Krippner and Friedman (eds) (2010), Chapter 6.
15. Ibid., Chapter 12. Several

meta-analyses showed there was a highly significant effect, with the exception of a skeptical paper by Milton and Wiseman (1999), which omitted a set of positive results that changed the overall balance to a significant positive effect (Milton, 1999). Also, Milton and Wiseman used a flawed method of analysis that failed to take into account each study's sample size. When their data were reanalysed correcting for this flaw, the overall effect was positive and statistically significant (Radin, in Krippner and Friedman (eds) (2010), Chapter 7).

16. Dalton (1997); Broughton and Alexander (1997).

17. My wife found this book in a second-hand bookshop. She immediately realised that it would interest me, which it certainly did, and bought it for me. This book has now been reprinted and is available again: see Long (2005).

18. Sheldrake (1999a), Chapter 3.

19. Sheldrake and Smart (1998).

20. Sheldrake and Smart (2000a).

21. After an experiment with Jaytee was shown on British television, several skeptics challenged Jaytee's ability to know when Pam was going home, and tried to explain them away. I invited one of them, Richard Wiseman, a conjurer, psychologist and member of the American Committee for the Scientific Investigation of Claims of the Paranormal (CSICOP), to carry out his own tests with Jaytee. Wiseman accepted my invitation and Pam and her family kindly helped him. In his tests, his assistant accompanied Pam all the time she was away from home, and told her when to return at a randomly selected time. Wiseman was with Jaytee, filming him. The results were very similar to my own tests; in fact the effect was even larger. In Wiseman's tests, Jaytee was at the window four per cent of the time in the main period of Pam's absence, and 78 per cent when she was on the way home (Sheldrake and Smart, 2000a). Wiseman and his colleague Matthew Smith, however, claimed that Jaytee had failed the test because he went to the window before Pam actually set off, and disregarded their own data showing that his waiting behaviour was very similar to that in my own tests (Wiseman et al., 1998). I replied to their claims (Sheldrake, 1999b) and there were two further rounds of response (Wiseman et al.,

2000, and Sheldrake, 2000).
For a summary of this controversy, see Carter (2010) and
Sheldrake (2011a). Wiseman
now concedes that his results
did in fact replicate my own,
saying, 'The patterning in my
studies is the same as the
patterning in Rupert's studies.'
(http://www.skeptiko.com/11-
dr-richard-wiseman-on-
rupert-sheldrakes-dogsthat-
know/)

22. Sheldrake and Smart (2000b).
23. Sheldrake (2011a).
24. Sheldrake and Morgana (2003).
25. Van der Post (1962), pp. 236–7
26. There were more female
    than male respondents,
    which is why the average, 92
    per cent, was not the mean
    of 96 and 85 per cent
    (Sheldrake, 2003a).
27. Lobach and Bierman (2004);
    Schmidt *et al.* (2009).
28. Sheldrake and Smart (2000a, b).
29. Sheldrake (2003a).
30. Radin (2007).
31. Einstein, in Einstein and
    Born (1971).
32. Sheldrake and Smart (2005),
    Sheldrake and Avraamides
    (2009); Sheldrake, Avraamides
    and Novak (2009).
33. Sheldrake and Lambert
    (2007); Sheldrake and
    Beeharee (2009).
34. See the online experiments
    portal at www.sheldrake.org

35. Sheldrake (2003a, 2011a).
36. Grant and Halliday (2010).
37. Sheldrake (2005c).
38. Sheldrake (2003a).
39. Sheldrake (2011a).
40. Sheldrake (2003a, 2011a).
41. Saltmarsh (1938).
42. Ibid.
43. Dunne (1927).
44. Radin (1997).
45. Radin (1997), Chapter 7.
46. Bierman and Scholte (2002);
    Bierman and Ditzhuijzen
    (2006); Bem (2011).
47. For example, Richard
    Wiseman, a well-known British
    skeptic, conceded that the data
    from experiments on ESP 'meet
    the usual standards for a
    normal claim, but are not
    convincing enough for an
    extraordinary claim'. http://
    subversivethinking.blogspot.
    com/2010/04/richard-wiseman-
    evidence-for-esp-meets.html
48. For well-informed discus-
    sions of skeptics' attitudes,
    see Griffin (2000), Chapter
    7; also Carter (2007) and
    McLuhan (2010).
49. French, in Henry (ed.)
    (2005), Chapter 5.
50. For skeptical discussion of
    skeptical claims, see www.
    skepticalinvestigations.org
51. See the Appendix to
    Sheldrake (2011a) and the
    Controversies section of my
    website, www.sheldrake.org.

52. Whitfield (2004).
53. http://www.skepticalinvesti-gations.org/New/Audio/telepathy.html
54. http://www.skepticalinvesti-gations.org/controversies/Euroskep_2005.htm

10: Is Mechanistic Medicine the Only Kind that Really Works?

1. Sheldrake (2009), Chapter 1.
2. Jones and Dangl (2006).
3. Elgert (2009).
4. Ibid.
5. Le Fanu (2000).
6. Ibid.
7. Ibid., pp. 177–8.
8. Weil (2004).
9. Ibid.
10. Le Fanu (2000).
11. Boseley (2002).
12. Goldacre (2010).
13. Ibid.
14. Wikipedia entry on 'Pharmaceutical lobby', http://en.wikipedia.org/wiki/Pharmaceutical_lobby
15. Goldacre (2009).
16. Stier (2010).
17. Ibid.
18. Mussachia (1995).
19. Rosenthal (1976).
20. Roberts et al. (1993).
21. Evans (2003).
22. Kirsch (2010).
23. Kirsch (2009).
24. Ibid.
25. Kaptchuck (1998).
26. Evans (2003).
27. Weil (2004).
28. Dossey (1991).
29. Moerman (2002).
30. Ibid.
31. Singh and Ernst (2009), p. 300.
32. Silverman (2009).
33. Reiche et al. (2005).
34. E. G., Pattie (1941); Stevenson (1997), p. 16.
35. Weil (2004), Chapter 21.
36. Freedman (1991).
37. Time (1952).
38. Mason (1955).
39. Weil (2004), Chapter 21.
40. Burns (1992).
41. Le Fanu (2000).
42. Source: US Centers for Disease Control and Prevention: http://www.cdc.gov/obesity/childhood/index.html
43. Kreitzer and Riff (2011).
44. Sheldrake (1999), Chapter 5.
45. Koenig (2008).
46. Ibid., Chapter 9.
47. Ibid., p. 143.
48. Crow (2011), p. 571.
49. Ibid.
50. Ibid.
51. UK Government (2010).
52. Le Fanu (2000), p. 400.
53. E.g., Singh and Ernst (2005).
54. World Health Organization (2003).
55. Singh and Ernst (2005).

56. Ibid.
57. Moncrieff (2009).
58. Kirsch (2009), p. 158.
59. Gray (2011), Chapter 2.
60. Ibid.
61. http://outthere.whatitcosts. com/cryogen-frozen.htm
62. Quoted by Willis (2009).
63. Hamilton (2005).
64. Source: American Medical Association: http://www. ama-assn.org/amednews/ 2009/08/24/prsa0824.htm
65. Zhang *et al.* (2009).
66. Temel *et al.* (2010).

11: Illusions of Objectivity

1. Lear (1965), p. 89.
2. Ibid., p. 114.
3. Meri (2005), pp. 138–9.
4. Lear (1965), pp. 103–4.
5. Ibid., Introduction.
6. Descartes (translation, 1985), Vol. 1, p. 127.
7. Zajonc (1993).
8. D'Espagnat (1976), p. 286.
9. Latour (2009), pp. 10–11.
10. Latour (1987); Collins and Pinch (1998).
11. Collins and Pinch (1998), p. 111.
12. Ibid., p. 42.
13. See the discussion in Chapter 4.
14. Sheldrake (2001).
15. Sheldrake (2004a).
16. Sheldrake (2001).
17. Alistair Cuthbertson (personal communication, 13 November 2010) carried out this survey of thirty-three heads of science from state schools at the Prince of Wales' Teaching Institute for science teachers in November 2010.
18. Sheldrake (2004a).
19. Medawar (1990).
20. Rosenthal (1976).
21. Rosenthal (1976), Chapter 10.
22. Sheldrake (1999c).
23. Watt and Nagtegaal (2004).
24. Sheldrake (1998b).
25. Sheldrake (1999c).
26. Sheldrake (1994), Chapter 7.
27. Sheldrake (1998b).
28. Sheldrake (1998c).
29. Sheldrake (1999d).
30. Wiseman and Watt (1999).
31. Enz (2009).
32. For example, Radin (2007) in his meta-analyses calculated how many unpublished negative sets of data would be needed to offset positive published results, and found that file drawer effects could not plausibly explain the overall positive results in parapsychological research.
33. Goldacre (2011).
34. Wicherts *et al.* (2006).
35. Broad and Wade (1985).
36. Ibid.
37. Dennett (2006).
38. Wikipedia entry on 'Schön scandal': http://en.wikipedia. org/wiki/Schön_scandal
39. *Nature* (2010).

40. *Daily Telegraph* (2010).
41. Broad and Wade (1985), pp. 141–2.
42. Hettinger (2010).
43. Oreskes and Conway (2010), Chapter 1.
44. Michaels (2005).
45. Ibid.
46. Oreskes and Conway (2010).
47. For a stimulating discussion of these questions, see Latour (2009), Chapter 3.

12: Scientific Futures

1. Royal Society (2011).
2. Ibid.
3. US Census data, accessed June 2011: http://www.census. gov/compendia/statab/2011/ tables/11s0807.pdf
4. Data for 2005: Royal Society (2005).
5. National Science Foundation: http://www.nsf.gov/statistics/ infbrief/nsf09314/
6. Smolin (2006).
7. Ziman (2003).
8. Feyerabend (2010).
9. http://www.cam.ac.uk/ admissions/undergraduate/ courses/natsci/part1b.html, accessed June 2011.
10. Fara (2009), pp. 191–6.
11. Ibid., pp. 194, 196.
12. Including some published only online. Source: Royal Society (2011).
13. Carr (2007).

14. Brooke (1991), p. 155.
15. Fara (2009), p. 197.
16. Ibid., p. xv.
17. Krönig (1992), p. 155.
18. Gervais (2010).
19. I have been fortunate to take part in many scientific and philosophical dialogues, which have been some of the most mind-opening experiences of my life. To name but a few, I discussed how modern physics might be related to morphogenetic fields with the quantum physicists David Bohm and Hans-Peter Dürr. The theologian Matthew Fox and I explored new connections between science and spirituality; some of our discussions were published in our books *Natural Grace* (1996) and *The Physics of Angels* (1996). In a series of annual dialogues, Andrew Weil and I discussed connections between scientific research, integrative medicine and consciousness studies, all available online in streaming audio on my website, www. sheldrake.org. In a series of three-way discussions, or trialogues, spread out over fifteen years, Ralph Abraham, a pioneering mathematician in the field of chaos theory, Terence McKenna, a researcher on the shamanic

use of psychedelic plants, and I explored a wide range of topics. Some of our trialogues were published in our books *Chaos, Creativity and Cosmic Consciousness* (2001) and *The Evolutionary Mind* (2005), and most are available online in streaming audio on my website.

20. UK Office of Science and Technology (2000).

21. Hansen (2010).

22. For example, the AIDS Treatment Activist Coalition: http://www.atac-usa.org/

23. Akrich *et al.* (2008).

24. Source: The Association of Medical Research Charities: http://www.amrc.org.uk/ our-members_member-profiles

25. Sample (2010).

26. I discussed this idea with leading politicians in Britain, both within the government and the opposition, and found that most were receptive to this possibility. I published it in *Nature* (Sheldrake, 2004b) and the *New York Times* (Sheldrake, 2003c), and it was taken up by the policy think tanks Demos (Wilsdon *et al.*, 2005). But nothing actually happened; it was simpler to leave things as they were, and changes in science funding systems are not big vote-winners. But this is still an open possibility.

27. Shannon (2002).

28. Viveiros de Castro (2004).

29. Wallace (2009), pp. 24–5.

30. Horgan (2003).

31. Weil (2004).

32. Heglund *et al.* (1995).

33. See for example my own explorations with the theologian Matthew Fox in Sheldrake and Fox (1996) and Fox and Sheldrake (1996).

34. Tarnas (1991), Chapter 3.

35. Suzuki (1998).

36. Royal Society (2011).

References

Adam, D. (2002), 'Flickering light raises possibility of changing "constant" ' *Nature*, 412, 757.

Aharonov, Y., Popescu, S., and Tollaksen, J. (2010), 'A time-symmetric formulation of quantum mechanics', *Physics Today*, November, 27–32.

Agnew, Z. K., Bhakoo, K. K., and Puri, B. K. (2007), 'The human mirror system: a motor resonance theory of mind-reading', *Brain Research Reviews*, 54, 286–93.

Akrich, M., Nunes, J., Paterson, F., and Rabeharisoa, V. (2008), *The Dynamics of Patient Organizations in Europe*, Presses des Mines, Paris.

Anfinsen, C. B., and Scheraga, H. A. (1975), 'Experimental and theoretical aspects of protein folding', *Advances in Protein Chemistry*, 29, 205–300.

Anway, M. D., Cupp, A. S., Uzumcu, M., and Skinner, M. K. (2005), 'Epigenetic transgenerational actions of endocrine disruptors and male fertility', *Science*, 308, 1466–9.

Auden, W. H. (2009), *The Selected Writings of Sydney Smith*, Faber & Faber, London.

Bacon, F. (1951), *The Advancement of Learning* and *New Atlantis*, Oxford University Press, London.

Banks, R. D., Blake, C. C. F., Evans, P. R., Haser, R., Rice, D. W., Hardy, G. W., Merrett, M., and Phillips, A. W. (1979), 'Sequence, structure and activities of phosphoglycerate kinase', *Nature*, 279, 773–7.

Barnett, S. A. (1981), *Modern Ethology*, Oxford University Press, Oxford.

Barrett, W. (1904), Address by the President, *Proceedings of the Society for Psychical Research*, 18, 323–50.

Barrow, J. D., and Webb, J. K. (2005), 'Inconstant constants: Do the inner workings of nature change with time?', *Scientific American*, June, 32–9.

Bekenstein, J. (2004), 'Relativistic gravitation theory for the modified Newtonian dynamics paradigm', *Physical Review D*, 70, Issue 8, 083509.

Belokov, A. V., and Hooper, D. (2010), 'Contribution of inverse Compton scattering to the diffuse extragalactic gamma-ray background from annihilating dark matter', *Physical Review D*, 81, 043505.

Bem, D. (2011), 'Feeling the future: experimental evidence for anomalous retroactive influences on cognition and affect', *Journal of Personality and Social Psychology*, 100, 407–25.

Bergson, H. (1911), *Creative Evolution*, Macmillan, London.

Bergson, H. (1946), *The Creative Mind*, Philosophical Library, New York.

Bernstein, J. (2002), *Polymorphism in Molecular Crystals*, Clarendon Press, Oxford.

Bierman, D. and Ditzhuijzen, J. (2006), 'Anomalous slow cortical components in a slot-machine task', *Proceedings of the 49th Annual Parapsychological Association*, 5–19.

Bierman, D., and Scholte, H. (2002), 'Anomalous anticipatory brain activation preceding exposure of emotional and neutral pictures', *Journal of International Society of Life Information Science*, 380–88.

Birge, W. T. (1929), 'Probable valves of the general physical constants' *Reviews of Modern Physics*, 33, 233–9.

Blackiston, D. J., Casey, E. S., and Weiss, M. R. (2008), 'Retention of memory through metamorphosis: Can a moth remember what it learned as a caterpillar?' *PLoS ONE*, 3 (3), e1736.

Blackmore, S. (1999), *The Meme Machine*, Oxford University Press, Oxford.

Blackmore, S. (2005), *Conversations on Consciousness*, Oxford University Press, Oxford.

Boakes, R. (1984), *From Darwin to Behaviourism*, Cambridge University Press, Cambridge.

Bohm, D. (1980), *Wholeness and the Implicate Order*, Routledge & Kegan Paul, London.

Bojowald, M. (2008), 'Big bang or big bounce? New theory on the universe's birth', *Scientific American*, October.

Boseley, S. (2002), 'Scandal of scientists who take money for papers ghostwritten by drug companies', *Guardian*, 7 February.

Bowler, P. J. (1984), *Evolution: The History of an Idea*, University of California Press, Berkeley.

Boycott, B. B. (1965), 'Learning in the octopus', *Scientific American*, 212 (3), 42–50.

Braud, W., Shafer, D., and Andrews, S. (1990), 'Electrodermal correlates of remote attention: Autonomic reactions to an unseen gaze', *Proceedings of Presented Papers, Parapsychology Association 33rd Annual Convention*, Chevy Chase, MD, 14–28.

Broad, W., and Wade, N. (1985), *Betrayers of the Truth: Fraud and Deceit in Science*, Oxford University Press, Oxford.

Brooke, J. H. (1991), *Science and Religion: Some Historical Perspectives*, Cambridge University Press, Cambridge.

Brooks, M. (2009), *13 Things That Don't Make Sense*, Profile Books, London.

Brooks, M. (2010), 'Operation alpha', *New Scientist*, 23 October, 33–5.

Broughton, R. S. and Alexander, C. M. (1997) Auroganzfeld II. An attempted replication of the PRL research', *Journal of Parapsychology*, 61, 209–226.

Brown, R. E. Fitzmyer, J. A., and Murphy, R. E. (1968), *The Jerome Bible Commentary*, Prentice-Hall, Englewood Cliffs, NJ.

Burnet, J. (1930) *Early Greek Philosophy*, A&C Black, London.

Burns, D. A. (1992), ' "Warts and all" – the history and folklore of warts: a review', *Journal of the Royal Society of Medicine*, 85, 37–40.

Bursen, H. A. (1978), *Dismantling the Memory Machine*, Reidel, Dordrecht.

Burtt, E. A. (1932), *The Metaphysical Foundations of Modern Physical Science*, Kegan Paul, Trench & Trubner, London.

Capra, F. (1996), *The Web of Life: A New Synthesis of Mind and Matter*, HarperCollins, London.

Carr, B. (ed.) (2007), *Universe or Multiverse?* Cambridge University Press, Cambridge.

Carr, B. (2008), 'Worlds apart? Can psychical research bridge the gap between matter and minds?', *Proceedings of the Society for Psychical Research*, 59, 1–96.

Carroll, S. B. (2005), *Endless Forms Most Beautiful*, Quercus, London.

Carroll, S. B. Grenier, J. K., and Weatherbee, S. D. (2001), *From DNA to Diversity: Molecular Genetics and the Evolution of Animal Design'*, Blackwell, Oxford.

Carter, C. (2007) *Parapsychology and the Skeptics*, Sterling House, Pittsburgh, PA.

Carter, C. (2010), ' "Heads I lose, Tails you win", or, How Richard

Wiseman nullifies positive results and what to do about it', *Journal of the Society for Psychical Research*, 74, 156–67.

Chivers, T. (2010), 'Neuroscience, free will and determinism: "I'm just a machine" ', *Daily Telegraph*, 12 October.

Cipolla-Neto, J., Horn, G., and McCabe, B. J. (1982), 'Hemispheric asymmetry and imprinting: the effect of sequential lesions to the Hyperstriatum ventrale', *Experimental Brain Research*, 48, 22–7.

Cohen, E. R., and Taylor, B. N. (1973), 'The 1973 least-squares adjustment of the fundamental constants', *Journal of Physical and Chemical Reference Data*, 2, 663–735.

Cohen, E. R., and Taylor, B. N. (1986), 'The 1986 CODATA recommended values of the fundamental physical constants', *Journal of Physical and Chemical Reference Data*, 17, 1795–803.

Cohn, N. (1957), *The Pursuit of the Millennium*, Secker & Warburg, London.

Cole, F. J. (1930), *Early Theories of Sexual Generation*, Clarendon Press, Oxford.

Coleman, W. (1977), *Biology in the Nineteenth Century*, Cambridge University Press, Cambridge.

Collins, H., and Pinch, T. (1998), *The Golem: What You Should Know About Science*, 2nd ed., Cambridge University Press, Cambridge.

Collins, J. (1965), *A History of Modern European Philosophy*, Bruce Publishing, Milwaukee, WI.

Conniff, R. (2006), 'Animal instincts', *Guardian*, 27 May.

Connor, S. (2011), 'For the love of God: Scientists in uproar at £1 million religion prize', *Independent*, 7 April.

Cooper, D., and Goodenough, L. (2010), 'Dark matter annihilation in the galactic center as seen by the Fermi gamma ray space telescope', http://arxiv.org/abs/1010.2752

Coopersmith, J. (2010), *Energy. The Subtle Concept: The Discovery of Feynman's Blocks from Leibniz to Einstein*, Oxford University Press, Oxford.

Corbett, J. (1986), *Jim Corbett's India*, Oxford University Press, Oxford.

Cottrell, J. E., Winer, G. A., and Smith, M. C. (1996), 'Beliefs of children and adults about feeling stares of unseen others', *Developmental Psychology*, 32, 50–61.

Cramer, J. (1986), 'The transactional interpretation of quantum mechanics', *Reviews of Modern Physics*, 58, 647–88.

Crick, F. (1966), *Of Molecules and Men*, University of Washington Press, Seattle.

Crick, F. (1984), 'Memory and molecular turnover', *Nature*, 312, 101.

Crick, F. (1994), *The Astonishing Hypothesis: The Scientific Search for the Soul*, Simon & Schuster, London.

Crow, M. M. (2011), 'Time to rethink the NIH', *Nature*, 471, 569–71.

Culotta, E. (2005) Chimp genome catalogs differences with humans. *Science*, 309, 1468–9.

Cunningham, A., and Jardine, N. (eds) (1990), *Romanticism and the Sciences*, Cambridge University Press, Cambridge.

Daily Telegraph (2010), ' "Atheists just as ethical as churchgoers", new research shows', *Daily Telegraph*, 9 February.

Dalton, K. (1997), 'Exploring the links: creativity and psi in the ganzfeld', *Proceedings of the Parapsychological Association 40th Annual Convention*, 119–31.

Danckwerts, P. V. (1982), Letter, *New Scientist*, 11 November, 380–81.

Darwin, C. (1859), *The Origin of Species*, Murray, London.

Darwin, C. (1875), *The Variation of Animals and Plants Under Domestication*, Murray, London.

Darwin, E. (1794–6; reprinted 1974), *Zoonomia*, 2 vols, AMS Press, New York.

Dasgupta, M. (2010), 'DIPAS concludes observational study on "Mataji" ', *Hindu*, 10 May.

Davies, P. (1984), *Superforce*, Heinemann, London.

Davies. P. (2006), *The Goldilocks Enigma: Why is the Universe Just Right For Life?*, Allen Lane, London.

Dawkins, R. (1976), *The Selfish Gene*, Oxford University Press, Oxford.

Dawkins, R. (1982), *The Extended Phenotype*, Oxford University Press, Oxford.

Dawkins, R. (2006), *The God Delusion*, Bantam, London.

De Bray, E. J. C. (1934), 'Velocity of light', *Nature*, 133, 948.

Dembski, W. (1998), *The Design Inference*, Cambridge University Press, Cambridge.

Dennett, D. (1991), *Consciousness Explained*, Little, Brown, Boston.

Dennett, D. (2006), *Breaking the Spell: Religion as a Natural Phenomenon*, Viking, New York, NY.

D'Espagnat, B. (1976), *Conceptual Foundations of Quantum Mechanics*, Benjamin, Reading, MA.

De Quincey, C. (2008), 'Reality bubbles', *Journal of Consciousness Studies*, 15, 94–101.

Descartes, R. (1985), *The Philosophical Writings of Descartes*, Cambridge University Press, Cambridge.

Dijksterhuis, E. J. (1961), *The Mechanization of the World Picture*, Oxford University Press, Oxford.

Dossey, L. (1991), *Meaning and Medicine*, Bantam Books, New York.

Driesch, H. (1914), *The History and Theory of Vitalism*, Macmillan, London.

Dubos, R. (1960), *Pasteur and Modern Science*, Anchor Books, New York.

Duncan, T., and Kennett, H. (2001), *GCSE Physics*, Murray, London.

Dunne, J. W. (1927), *An Experiment With Time*, Faber & Faber, London.

Dürr, H-P., and Gottwald, F-T. (eds) (1997), *Rupert Sheldrake in der Diskussion: Das Wagnis einer neuen Wissenschaft des Lebens*, Scherz Verlag, Bern.

Dyson, F. (1979), *Disturbing the Universe*, Harper & Row, New York.

Einstein, A., and Born, M. (1971), *The Born-Einstein Letters*, Walker, New York.

Elgert, K. D. (2009), *Immunology: Understanding the Immune System*, Wiley, Hoboken, NJ.

Ellis, G. (2011), 'The untestable multiverse', *Nature*, 469, 295–295.

Elsasser, W. M. (1975), *The Chief Abstractions of Biology*, North Holland, Amsterdam.

Enz, C. P. (2009), 'Rational and irrational features in Wolfgang Pauli's life', in *Of Matter and Spirit: Selected Essays by Charles P. Enz*, World Scientific, Hackensack, NJ.

Evans, D. (2003), *Placebo: The Belief Effect*, HarperCollins, London.

Fara, P. (2009), *Science: A Four Thousand Year History*, Oxford University Press, Oxford.

Feyerabend, P. (2010), *Against Method*, 4th ed., Verso, London.

Feynman, R. (1962), *Quantum Electrodynamics*, Addison-Wesley, Reading, MA.

Feynman, R. (1964), *The Feynman Lectures on Physics, Vol. 1*, Addison-Wesley, Reading, MA.

Filippini, G., and Gramaccioli, C. M. (1989), 'Benzene crystals at low temperature: A harmonic lattice-dynamical calculation', *Acta Crystallographica*, A45, 261–261.

Flew, A. (ed.) (1979), *A Dictionary of Philosophy*, Macmillan, London.

Flynn, J. (2007), *What is Intelligence?*, Cambridge University Press, Cambridge.

Forster, J. R. (1778), *Observations Made During a Voyage Around the World*, Robinson, London.

Fox, M., and Sheldrake, R. (1996), *The Physics of Angels: Exploring the Realm Where Science and Spirit Meet*, Harper, San Francisco.

Frankenfield, D. C. (2010), 'On heat, respiration and calorimetry', *Nutrition*, 26, 939–50.

Freedman, R. R. (1991), 'Physiological mechanisms of temperature biofeedback', *Applied Psychophysiology and Biofeedback*, 16, 95–115.

Fröhlich, F., and McCormick, D. A. (2010), 'Endogenous electric fields may guide neocortical network activity', *Neuron*, 67, 129–43.

Galton, F. (1875), 'The history of twins as a criterion of the relative powers of nature and nurture', *Fraser's Magazine*, 12, 566–76.

Gardner, H. (1974), *The Shattered Mind*, Vintage Books, New York.

Gerhart, J., and Kirschner, M. (1997), *Cells, Embryos and Evolution*, Blackwell Science, Oxford.

Gershteyn, M. L., Gershteyn, L. I., Gershteyn, A., and Karagioz, O. V. (2002), 'Experimental evidence that the gravitational constant varies with orientation', http://arxiv.org/pdf/physics/0202058v2

Gervais, R. (2010), 'Why I'm an atheist', *Wall Street Journal*, 19 December.

Gibson, J. J. (1986), *The Ecological Approach to Visual Perception*, Lawrence Erlbaum Associates, Hillsdale, NJ.

Gilbert, W. (1600; reprinted 1991), *De Magnete*, Dover Books, New York.

Gilson, E. (1984), *From Aristotle to Darwin and Back Again*, University of Notre Dame Press, Notre Dame, IN.

Gleik, J. (1988), *Chaos: Making a New Science*, Heinemann, London.

Goho, A. (2004), 'The crystal form of a drug can be the secret of its success', *Science News*, 166, 122–4.

Goldacre, B. (2009), 'Dithering over statins': side-effects label finally ends', *Guardian*, 21 November.

Goldacre, B. (2010), 'Medical ghostwriters who build a brand', *Guardian*, 18 September.

Goldacre, B. (2011), 'Backwards step on looking into the future', *Guardian*, 23 April.

Goodwin, B. (1994), *How the Leopard Changed its Spots*, Weidenfeld & Nicolson, London.

Gould, S. J. (1989), *Wonderful Life: The Burgess Shale and the Nature of History*, Hutchinson, London.

Gould, S. J. (1999), *Rock of Ages: Science and Religion in the Fullness of Life*, Ballantine, New York.

Grant, R., and Halliday, T. (2010), 'Predicting the unpredictable: evidence of pre-seismic anticipatory behaviour in the common toad', *Journal of Zoology*, 281, 263–71.

Gray, Jeffrey (2004), *Consciousness: Creeping Up on the Hard Problem*, Oxford University Press, Oxford.

Gray, John (2002), *Straw Dogs: Thoughts on Humans and Other Animals*, Granta Books, London.

Gray, John (2007), *Black Mass: Apocalyptic Religion and the Death of Utopia*, Allen Lane, London.

Gray, John (2011) *The Immortalization Commission: The Strange Quest to Cheat Death*, Allen Lane, London.

Grayling, A. C. (2011) 'Psychology: how we form beliefs', *Nature* 474, 446–7.

Green, E. D., and Guyer, M. S. (2011), 'Charting a course for genomic medicine from base pairs to bedside', *Nature*, 470, 204–13.

Greene, B. (2011), *The Hidden Reality: Parallel Universes and the Deep Laws of the Cosmos*, Allen Lane, London.

Greenfield, S. (2000), *Brain Story: Unlocking Our Inner World of Emotions, Memories, Ideas and Desires*, BBC, London.

Griffin, D. R. (1998), *Unsnarling the World-Knot: Consciousness, Freedom and the Mind-Body Problem*, Wipf & Stock, Eugene, OR.

Griffin, D. R. (2000), *Religion and Scientific Naturalism: Overcoming the Conflicts*, State University of New York Press, Albany, NY.

Grundlach, J. H., and Merkowitz, S. M. (2000), 'Measurement of Newton's constant using a torsion balance with acceleration feedback', *Physical Review Letters*, 85, 2869–72.

Haemmerling, J. (1963), 'Nucleo-cytoplasmic interactions in Acetabularia and other cells', *Annual Reviews of Plant Physiology*, 14, 65–92.

Hamilton, C. (2005), 'Chasing immortality: the technology of eternal life', *What Is Enlightenment?*, 30, 16–19.

Hampshire, S. (1951), *Spinoza*, Penguin, Harmondsworth.

Hansen, J. (2010), *Biotechnology and Public Engagement in Europe*, Palgrave Macmillan, London.

Harman, P. M. (1982), *Energy, Force and Matter: The Conceptual Development of Nineteenth-Century Physics*, Cambridge University Press, Cambridge.

Hawking, S. (1988), *Is the End in Sight for Theoretical Physics?*, Cambridge University Press, Cambridge.

Hawking, S., and Mlodinow, L. (2010), *The Grand Design: New Answers to the Ultimate Questions of Life*, Bantam Press, London.

Hazen, R. (1989), 'Battle of the supermen', *Guardian*, 15 April.

Heglund, N. C., Willems, P. A., Penta, M., and Cavagna, G. A. (1995), 'Energy-saving gait mechanics with head-supported loads', *Nature*, 375, 52–4.

Henry, J. (ed.) (2005), *Parapsychology: Research on Exceptional Experiences*, Routledge, Hove.

Hettinger, T. P. (2010), 'Misconduct: don't assume science is self-correcting', *Nature*, 466, 1040.

Hinde, R. A. (1982), *Ethology*, Fontana, London.

Hodges, A. (1983), *Alan Turing: The Enigma of Intelligence*, Hutchinson, London.

Holden, A., and Singer, P. (1961), *Crystals and Crystal Growing*, Heinemann, London.

Holder, N. (1981), 'Regeneration and compensatory growth', *British Medical Bulletin*, 37, 227–32.

Holding, S. C., Stacey, F. D., and Tuck, G. J. (1986), 'Gravity in mines – an investigation of Newton's law', *Physics Review Letters D*, 33, 3487–94.

Horgan, J. (1997a), 'Get smart, take a test: A long term rise in IQ scores baffles intelligence experts', *Scientific American*, November, 10–11.

Horgan, J. (1997b), *The End of Science: Facing the Limits of Knowledge in the Twilight of the Scientific Age*, Little, Brown, London.

Horgan, J. (2003), *Rational Mysticism: Dispatches from the Border Between Science and Spirituality*, Houghton Mifflin, Boston.

Horn, G. (1986), *Memory, Imprinting and the Brain: An Inquiry into Mechanisms*, Clarendon Press, Oxford.

Howe, D., and Rhee, S. Y. (2008), 'The future of biocuration', *Nature*, 455, 47–8.

Hume, D. (2008), *Dialogues Concerning Natural Religion*, Oxford University Press, Oxford.

Humphrey, N. (2011), *Soul Dust: The Magic of Consciousness*, Quercus, London.

Hunter, I. M. L. (1964), *Memory*, Penguin, Harmondsworth.

Huxley, F. (1959), 'Charles Darwin: life and habit', *American Scholar* (Fall/Winter), 1–19.

Huxley, T. H. (1867), *Hardwicke's Science Gossip*, 3, 74.

Huxley, T. H. (1893), *Methods and Results*, Macmillan, London.

Iacono, W. G., and McGue, M. (2002), 'Minnesota Twin Family Study', *Twin Studies*, 5, 482–7.

Inge, W. R. (1929), *The Philosophy of Plotinus*, Longmans, London.

Jennings, H. S. (1906), *Behavior of the Lower Organisms*, Columbia University Press, New York.

Jones, J. D. G., and Dangl, J. L. (2006), 'The plant immune system', *Nature*, 444, 323–9.

Kahn, F. (1949), *The Secret of Life: The Human Machine and How It Works*, Odhams, London.

Kandel, E. R. (2003), 'The molecular biology of memory storage: a dialogue between genes and synapses', in Jornvall, H. (ed), *Nobel Lectures, Physiology or Medicine 1995–2000*, World Scientific, Singapore.

Kandel, E. R., Schwartz, J. H., and Jessell, T. M. (1995), *Essentials of Neuroscience and Behavior*, Appleton & Lang, Norwalk, CT.

Kaptchuk, T. J. (1998), 'Intentional ignorance: a history of blind assessment in medicine', *Bulletin of the History of Medicine*, 72, 389–443.

Kealey, T. (1996), *The Economic Laws of Scientific Research*, Macmillan, London.

Kekreja, L. M. (2009), 'Calls to counter science scepticism are irrelevant in India', *Nature*, 459, 321.

Khoury, M. J., Evans, J., and Burke, W. (2010), 'A reality check for personalized medicine', *Nature*, 464, 680.

Kiernan, V. (1995), 'Gravitational constant is up in the air', *New Scientist*, 29 April, 18.

Kirsch, I. (2009), *The Emperor's New Drugs: Exploding the Antidepressant Myth*, Bodley Head, London.

Kirsch, I. (2010), 'Not all placebos are born equal', *New Scientist*, 11 December, 30–33.

Klein, M., and Kandel, E. R. (1978), 'Presynaptic modulation of voltage-dependent Ca^{2+} current: mechanism for behavioral sensitization in *Aplysia californica*', *Proceedings of the National Academy of Sciences USA*, 75, 3512–16.

Koenig, H. (2008), *Medicine, Religion and Health: Where Science and*

Spirituality Meet, Templeton Foundation Press, West Conshohocken, PA.

Koestler, A. (1967), *The Ghost in the Machine*, Hutchinson, London.

Kreitzer, M. J., and Riff, K. (2011), 'Spirituality and heart health', in Devries, S., and Dalen, J. E. (eds), *Integrative Cardiology*, Oxford University Press, New York.

Kretzman, N., and Stump, E. (eds) (1993), *The Cambridge Companion to Aquinas*, Cambridge University Press, Cambridge.'

Krippner, S., and Friedman, H. L. (eds) (2010), *Debating Psychic Experience: Human Potential or Human Illusion*, Praeger, Santa Barbara, CA.

Krönig, J. (1992), *Spuren*, Zweitausendeins, Frankfurt.

Kuhn, T. S. (1959), 'Energy conservation as an example of simultaneous discovery', in Clagett, M. (ed.), *Critical problems in the History of Science*, University of Wisconsin Press, Madison, WI.

Kuhn, T. S. (1970), *The Structure of Scientific Revolutions*, 2nd ed., University of Chicago Press, Chicago.

Lamarck, J.-B. (1914), *Zoological Philosophy*, Macmillan, London.

Laplace, P. S. (1819; reprinted 1951), *A Philosophical Essay on Probabilities*, Dover, New York.

Lashley, K. S. (1929), *Brain Mechanisms and Intelligence*, Chicago University Press, Chicago.

Lashley, K. S. (1950), 'In search of the engram', *Symposium of the Society for Experimental Biology*, 4, 454–83.

Laszlo, E. (2007), *Science and the Akashic Field*, Inner Traditions, Rochester, VT.

Latham, J. (2011), 'The failure of the genome', *Guardian*, 18 April.

Latour, B. (1987), *Science in Action: How to Follow Scientists and Engineers Through Society*, Harvard University Press, Cambridge, MA.

Latour, B. (2009), *Politics of Nature: How to Bring the Sciences into Democracy*, Harvard University Press, Cambridge, MA.

Lear, J. (1965), *Kepler's Dream*, University of California Press, Berkeley.

Le Fanu, J. (2000), *The Rise and Fall of Modern Medicine*, Abacus, London.

Lehar, S. (1999), 'Gestalt isomorphism and the quantification of spatial perception', *Gestalt Theory*, 21, 122–39.

Lehar, S. (2004), 'Gestalt isomorphism and the primacy of subjective conscious experience', *Behavioral and Brain Sciences*, 26, 375–444.

Lewin, R. (1980), 'Is your brain really necessary?', *Science*, 210, 1232.

Libet, B. (1999), 'Do we have free will?', *Journal of Consciousness Studies*, 6, 47–57.

Libet, B. (2003), 'Can conscious experience affect brain activity?', *Journal of Consciousness Studies*, 10, 24–8.

Libet, B. (2006), 'Reflections on the interaction of the mind and brain', *Progress in Neurobiology*, 78, 322–6.

Libet, B., Elwood, W., Feinstein, B., and Pearl, D. K. (1979), 'Subjective referral of the timing for a conscious sensory experience', *Brain*, 102, 193–224.

Lightman, B. V. (2007), *Victorian Popularizers of Science: Designing Nature for New Audiences*, University of Chicago Press, Chicago.

Lindberg, D. C. (1981), *Theories of Vision from Al-Kindi to Kepler*, Chicago University Press, Chicago.

Lobach, E., and Bierman, D. J. (2004), 'Who's calling at this hour? Local sidereal time and telephone telepathy', in *Proceedings of the 47th Parapsychological Association Annual Convention* (pp. 91–7), Vienna.

Long, C. H. (1983), *Alpha: The Myths of Creation*, Oxford University Press, New York.

Long, W. (1919), *How Animals Talk*, Harper, New York.

Long, W. (2005, reprinted), *How Animals Talk*, Park Street Press, Rochester, VT.

Lorayne, H. (1950), *How to Develop a Super-Power Memory*, Thomas, Preston.

Lu, J., Tapia, J. C., White, O. L., and Lichtman, J. W. (2009), 'The interscutularis muscle connectome' *Public Library of Science Biology*, e 1000032. doi:10.1371/journal.pbio.1000032

Luria, A. R. (1970), 'The functional organization of the brain', *Scientific American*, 222(3), 66–78.

Luria, A. R. (1973), *The Working Brain*, Penguin, Harmondsworth.

Maddox, J. (1981), 'A book for burning?', *Nature*, 293, 245–6.

Malhotra, R., Holman, M., and Ito, T. (2001), 'Chaos and stability of the solar system', *Proceedings of the National Academy of Sciences US*, 98, 12342–3.

Manolio, T. A., Collins, F. S., and twenty-five others (2009), 'Finding the missing heritability of complex diseases', *Nature*, 461, 747–53.

Mason, A. A. (1955), 'Ichthyosis and hypnosis', *British Medical Journal*, 2 July, 57–8.

Mayr, E. (1982), *The Growth of Biological Thought*, Harvard University Press, Cambridge, MA.

McLuhan, R. (2010), *Randi's Prize: What Skeptics Say About the Paranormal, Why They Are Wrong and Why It Matters*, Matador, Leicester.

Medawar, P. B. (1990), *The Threat and the Glory: Reflections on Science and Scientists*, HarperCollins, London.

Medvedev, Z. A. (1969), *The Rise and Fall of T. D. Lysenko*, Columbia University Press, New York.

Meri, J. W. (2005) *Medieval Islam Civilization: An Encyclopedia*. Routledge, London.

Michaels, D. (2005), 'Doubt is their product', *Scientific American*, June.

Midgley, M. (2002), *Evolution As A Religion*, Routledge, London.

Milton, J. and Wiseman R. (1999), 'Does psi exist? Lack of replication of an anomalous process of information transfer', *Psychological Bulletin*, 125, 387–391.

Milton, J. (1999), 'Should ganzfeld research continue to be crucial in the search for a replicable psi effect?', *Journal of Parapsychology*, 63, 309–33.

Mitchell, M. (2009), *Complexity: A Guided Tour*, Oxford University Press, New York.

Moerman, D.E. (2002) Meoning, Medicine and the Place is Effect. Cambridge University Press, Cambridge.

Mohr, P. J., and Taylor, B. N. (2001), 'Adjusting the values of the fundamental constants', *Physics Today*, 54, 29.

Moncrieff, J. (2009), *The Myth of the Chemical Cure: A Critique of Psychiatric Drug Treatment*, Palgrave Macmillan, London.

Monod, J. (1972), *Chance and Necessity*, Collins, London.

Munowitz, M. (2005), *Knowing: The Nature of Physical Law*, Oxford University Press, Oxford.

Murphy, G., and Ballou, R.O. (eds) (1961), *William James on Psychical Research*, Chatto and Windus, London.

Mussachia, M. (1995), 'Objectivity and repeatability in science', *Skeptical Inquirer*, 19 (6), 33–5, 56.

National Science Board (2010), *Science and Engineering Indicators 2010*, National Science Foundation, Washington.

Nature (2010), 'News briefing', *Nature*, 467, 11.

Nature (2011), Editorial, 'Best is yet to come', *Nature* 470, 140.

Needham, J. (1959), *A History of Embryology*, Cambridge University Press, Cambridge.

Nemethy, G., and Scheraga, H. A. (1977), 'Protein folding', *Quarterly Review of Biophysics*, 10, 239–352.

Newton, I. (1704; reprinted 1952), *Opticks*, Dover Publications, New York.

Nietzsche, F. W. (1911), 'Eternal recurrence: the doctrine expounded and substantiated', in *The Complete Works of Friedrich Nietzsche*, Vol. 16, ed. O. Levy, Foulis, Edinburgh.

Noble, D. (2006), *The Music of Life: Biology Beyond the Genome*, Oxford University Press, Oxford.

Noë, A. (2009), *Out of Our Heads: Why You Are Not Your Brain, and Other Lessons from the Biology of Consciousness*, Hill & Wang, New York.

Nordenskiold, E. (1928), *The History of Biology*, Tudor, New York.

Olsen, M. V., and Varki, A. (2004), 'The chimpanzee genome – a bitter-sweet celebration', *Science*, 305, 191–2.

Oreskes, N. and Conway, E. K. (2010), *Merchants of Doubt: How a Handful of Scientists Obscured the Truth on Issues from Tobacco Smoke to Global Warming*, Bloomsbury Press, New York.

Ostriker, J. P., and Steinhardt, P. J. (2001), 'The quintessential universe', *Scientific American*, January, 46–53.

Pagels, H. R. (1983), *The Cosmic Code*, Michael Joseph, London.

Paley, W. (1802), *Natural Theology*, J. Vincent, Oxford.

Partridge, E. (1961), *Origins*, Routledge & Kegan Paul, London.

Pattie, F. (1941), 'The production of blisters by hypnotic suggestion: a review', *Journal of Abnormal and Social Psychology*, 36, 62–72.

Pauli, W., and Jung, C. G. (2001), *Atom and Archetype: The Pauli/Jung Letters 1932–1958*, Princeton University Press, Princeton.

Penfield, W. (1975), *The Mystery of the Mind*, Princeton University Press, Princeton.

Penfield, W., and Roberts L. (1959), *Speech and Brain Mechanisms*, Princeton University Press, Princeton.

Penrose, R. (2010), *Cycles of Time: An Extraordinary New View of the Universe*, Bodley Head, London.

Petley, B. W. (1985), *The Fundamental Physical Constants and the Frontiers of Metrology*, Adam Hilger, Bristol.

Petronis, A. (2010), 'Epigenetics as a unifying principle in the aetiology of complex traits and diseases', *Nature*, 465, 721–7.

Piaget, J. (1973), *The Child's Conception of the World*, Granada, London.

Pisano, G. P. (2006), *Science Business: The Promise, the Reality and the Future of Biotech*, Harvard Business School, Boston, MA.

Plato (2000, trans. B. Joiwett), *The Republic*, Dover Books, New York.

Plotinus, trans. MacKenna, S. (1956), *The Enneads*, Faber & Faber, London.

Popper, K. R., and Eccles. J. C. (1977), *The Self and Its Brain*, Springer International, Berlin.

Potters, V. G. (1967), *C. S. Peirce on Norms and Ideals*, University of Massachusetts, Worcester, MA.

Pribram, K. H. (1971), *Languages of the Brain*, Prentice Hall, Englewood Cliffs, NJ.

Pribram, K. H. (1979), 'Transcending the mind-brain problem', *Zygon*, 14, 103–24.

Qiu, J. (2006), 'Unfinished symphony', *Nature*, 441, 143–5.

Radin, D. (1997), *The Conscious Universe: The Scientific Truth of Psychic Phenomena*, HarperCollins, San Francisco.

Radin, D. (2007), *Entangled Minds: Extrasensory Experiences in a Quantum Reality*, Paraview Pocket Books, New York.

Recordon, E. G., Stratton, F. J. M., and Peters, R. A. (1968), 'Some trials in a case of alleged telepathy', *Journal of the Society for Psychical Research*, 44, 390–201.

Rees, M. (1997), *Before the Beginning: Our Universe and Others*, Simon & Schuster, London.

Rees, M. (2004), *Our Final Century: The 50/50 Threat to Humanity's Survival*, Arrow, London.

Reich, E. S. (2010), 'G-whizzes disagree over gravity', *Nature*, 466, 1030.

Reiche, E. M. V., Nunes, S. O. V., and Morimoto, H. K. (2005), 'Stress, depression, the immune system and cancer', *Lancet Oncology*, 5, 617–25.

Rizzolatti, G., Fadiga, L., Fogassi, L., and Gallese, V. (1999) 'Resonance behaviors and mirror neurons', *Archives Italiennes de Biologie*, 137, 85–100.

Roberts, A. H., Kewman, D. G., Mercier, L., and Hovell, H. (1993), 'The power of nonspecific effects in healing: implications for psychosocial and biological treatments', *Clinical Psychology Review*, 13, 375–91.

Robertson, B. E., Ellis, R. S., Dunlop, J. S., McLure, R. J., and Stark, D. P. (2010), 'Early star-forming galaxies and the reionization of the universe', *Nature*, 468, 49–55.

Rose, S. P. R. (1986), 'Memories and molecules', *New Scientist*, 112 (27 November), 40–44.

Rose, S. P. R., and Csillag, A. (1985), 'Passive avoidance training results in lasting changes in deoxyglucose metabolism in left hemisphere regions of chick brain', *Behavioural and Neural Biology*, 44, 315–24.

Rose, S. P. R., and Harding, S. (1984), 'Training increases 3H fucose incorporation in chick brain only if followed by memory storage', *Neuroscience*, 12, 663–7.

Rosenthal, R. (1976), *Experimenter Effects in Behavioral Research*, John Wiley, New York.

Royal Society (2005), *A Degree of Concern? UK First Degrees in Science, Technology and Mathematics*, Royal Society Policy Document 32/06, London.

Royal Society (2011), *Knowledge, Networks and Nations: Global Scientific Collaboration in the 21st Century*, Royal Society Policy Document 03/11, London.

Rubery, P. H. and Sheldrake, R. (1974) 'Carrier-mediated auxin transport', *Planta*, 118, 101–210.

Russell, E.S. (1945), *The Directiveness of Organic Activities*, Cambridge University Press, Cambridge.

Sacks, O. (1985), *The Man Who Mistook His Wife for a Hat*, Duckworth, London.

Saltmarsh, F. H. (1938), *Foreknowledge*, Bell, London.

Sample, I. (2010), 'Spending review spares science budget from deep cuts', *Guardian*, 19 October.

Sarton, G. (1955), Introductory essay, in J. Needham, ed., *Science, Religion and Reality*, Braziller, New York.

Satprem (2000), *Sri Aurobindo or the Adventure of Consciousness*, Mira Aditi Centre, Mysore.

Schelling, F. von (1988), *Ideas for a Philosophy of Nature*, Cambridge University Press, Cambridge.

Schmidt, S., Erath, D., Ivanova, V., and Walach, H. (2009), 'Do you know who is calling? Experiments on anomalous cognition in phone call receivers', *Open Psychology Journal*, 2, 12–18.

Schmidt, S., Schneider, R., Utts, J., and Walach, H. (2004), 'Distant intentionality and the feeling of being stared at: Two meta-analyses', *British Journal of Psychology*, 95, 235–47.

Schnabel, U. (2009), 'Ein Portwein auf die Gene', *Die Zeit*, 9 July.

Schwarz, J. P., Robertson, D. S., Niebauer, T. M., and Fuller, J. E. (1998), 'A free-fall determination of the Newtonian constant of gravity', *Science*, 282, 2230–34.

Searle, J. (1992), *The Rediscovery of the Mind*, MIT, Cambridge, MA.

Searle, J. (1997), 'Consciousness and the philosophers', *New York Review of Books*, 6 March, 43–50.

Shannon, B. (2002), *Antipodes of the Mind: Charting the Phenomenology of the Ayahuasca Experience*, Oxford University Press, Oxford.

Sheldrake, R. (1973)? 'The production of hormones in higher plants', *Biological Reviews*, 48, 509–99.

Sheldrake, R. (1974), 'The ageing death cells' growth', *Nature*, 250, 381–50.

Sheldrake, R. (1981; second ed. 1985); *A New Science of Life: The Hypothesis of Formative Causation*, Blond & Briggs, London.

Sheldrake, R. (1984), 'Pigeon pea physiology', in *The Physiology of Tropical Crops* (ed. P. H. Goldsworthy), Blackwell, Oxford.

Sheldrake, R. (1987), 'A perennial cropping system for pigeonpea grown in post-rainy season', *Indian Journal of Agricultural Sciences*, 57, 895–9.

Sheldrake, R. (1988a), *The Presence of the Past: Morphic Resonance and the Habits of Nature*, Collins, London.

Sheldrake, R. (1988b), 'Cattle fooled by phoney grids', *New Scientist*, 11 February, 65.

Sheldrake, R. (1990), *The Rebirth of Nature: The Greening of Science and God*, Century, London.

Sheldrake, R. (1992a), 'An experimental test of the hypothesis of formative causation', *Biology Forum*, 85, 431–43.

Sheldrake, R. (1992b), 'Rose refuted', *Biology Forum*, 85, 455–60.

Sheldrake, R. (1994), *Seven Experiments That Could Change the World: A Do-It-Yourself Guide to Revolutionary Science*, Fourth Estate, London.

Sheldrake, R. (1998a), 'Perceptive pets with puzzling powers: three surveys', *International Society for Anthrozoology Newsletter*, 15, 2–5.

Sheldrake, R. (1998b), 'Experimenter effects in scientific research: how widely are they neglected?' *Journal of Scientific Exploration*, 12, 73–8.

Sheldrake, R. (1998c), 'Could experimenter effects occur in the physical and biological sciences?', *Skeptical Inquirer*, 22, 57–8.

Sheldrake, R. (1999a), *Dogs That Know When Their Owners Are Coming Home, and Other Unexplained Powers of Animals*, Hutchinson, London.

Sheldrake, R. (1999b), 'Commentary on a paper by Wiseman, Smith and Milton on the "psychic pet" phenomenon', *Journal of the Society for Psychical Research*, 63, 306–11.

Sheldrake, R. (1999c), 'How widely is blind assessment used in scientific research?' *Alternative Therapies*, 5, 88–91.

Sheldrake, R. (1999d), 'Blind belief', *Skeptic*, 12 (2), 7–8.

Sheldrake, R. (2000), 'The "psychic pet" phenomenon', *Journal of the Society for Psychical Research*, 64, 126–8.

Sheldrake, R. (2001), 'Personally speaking', *New Scientist*, 19 July.

Sheldrake, R. (2003a), *The Sense of Being Stared At, and Other Aspects of the Extended Mind*, Crown, New York.

Sheldrake, R. (2003b), 'Set them free', *New Scientist*, 19 April.

Sheldrake, R. (2003c), 'Really popular science', *New York Times*, 4 January.

Sheldrake, R. (2004a), 'Are we active? Or should the passive be used?' *School Science Review*, 86, 8–10.

Sheldrake, R. (2004b), 'Public participation: let the public pick projects', *Nature*, 432, 271.

Sheldrake, R. (2005a), 'The sense of being stared at. Part 1. Is it real or illusory?', *Journal of Consciousness Studies*, 12, 10–31.

Sheldrake, R. (2005a), 'The sense of being stared at Part 2. Its implications for theories of vision', *Journal of Consciousness Studies*, 12, 32–49.

Sheldrake (2005c), 'Why did so many animals escape December's tsunami?', *Ecologist*, March.

Sheldrake, R. (2009), *A New Science of Life* (3rd ed.), Icon Books, London.

Sheldrake, R. (2011a), *Dogs That Know When Their Owners Are Coming Home, and Other Unexplained Powers of Animals* (2nd ed.), Three Rivers Press, New York.

Sheldrake, R. (2011b), *The Presence of the Past: Morphic Resonance and the Habits of Nature* (2nd ed.), Icon Books, London.

Sheldrake, R., and Avraamides, L. (2009), 'An automated test for telepathy in connection with emails', *Journal of Scientific Exploration*, 23, 29–36.

Sheldrake, R., Avraamides, L., and Novak, M. (2009), 'Sensing the sending of SMS messages: An automated test', *Explore: The Journal of Science and Healing*, 5, 272–6.

Sheldrake, R., and Beeharee, A. (2009), 'A rapid online telepathy test', *Psychological Perspectives*, 104, 957–70.

Sheldrake, R., and Fox, M. (1996), *Natural Grace: Dialogues on Science and Spirituality*, Bloomsbury, London.

Sheldrake, R., and Lambert, M. (2007), 'An automated online telepathy test', *Journal of Scientific Exploration*, 21, 511–22.

Sheldrake, R., McKenna, T., and Abraham, R. (2002), *Chaos, Creativity and Cosmic Consciousness*, Part Street Press, Rochester, VT.

Sheldrake, R., McKenna, T., and Abraham, R. (2005), *The Evolutionary Mind: Conversations on Science, Imagination and Spirit*, Monkfish Books, Rhinebeck, NY.

Sheldrake, R., and Moir, G. F. J. (1970), 'A cellulase in *Hevea* later', *Physiologia Plantarum*.

Sheldrake, R., and Morgana, A. (2003), 'Testing a language-using parrot for telepathy', *Journal of Scientific Exploration*, 17, 601–15.

Sheldrake, R., and Smart, P. (1998), 'A dog that seems to know when his owner is returning: Preliminary investigations', *Journal of the Society for Psychical Research*, 62, 220–32.

Sheldrake, R., and Smart, P. (2000a), 'A dog that seems to know when his owner is coming home: Videotaped experiments and observations', *Journal of Scientific Exploration*, 14, 233–55.

Sheldrake, R., and Smart, P. (2000b), 'Testing a return-anticipating dog, Kane', *Anthrozoos*, 13, 203–12.

Sheldrake, R., and Smart, P. (2003a), 'Experimental tests for telephone telepathy', *Journal of the Society for Psychical Research*, 67, 174–99.

Sheldrake, R., and Smart, P. (2003b), 'Videotaped experiments on telephone telepathy', *Journal of Parapsychology*, 67, 147–66.

Sheldrake, R., and Smart, P. (2005), 'Testing for telepathy in connection with e-mails', *Perceptual and Motor Skills*, 101, 771–86.

Shermer, M. (2011), *The Believing Brain: From Ghosts and Gods to Politics and Conspiracies – How We Construct Beliefs and Reinforce them as Truths*, Times Books, New York.

Silverman, S, (2009), 'Placebos are getting more effective. Drugmakers are desperate to know why', *Wired Magazine*, 24 August.

Sinclair, U. (1930), *Mental Radio*, Werner Laurie, London.

Singh, S. (2004), *Big Bang*, Fourth Estate, London.

Singh, S., and Ernst, E. (2009), *Trick or Treatment? Alternative Medicine on Trial*, Corgi Books, London.

Skrbina, D. (2003), 'Panpsychism as an underlying theme in Western philosophy', *Journal of Consciousness Studies*, 10, 4–46.

Smith, A. P. (1978), 'An investigation of the mechanisms underlying nest construction in the mud wasp Paralastor sp.', *Animal Behaviour*, 26, 232–40.

Smolin, L. (2006), *The Trouble With Physics: The Rise of String Theory, The Fall of a Science, and What Comes Next*, Allen Lane, London.

Smolin, L. (2010), 'Space-time turnaround', *Nature*, 467, 1034–5.

Smuts, J. C. (1926), *Holism and Evolution*, Macmillan, London.

Sobel, D. (1998), *Longitude: The True Story of a Scientific Genius Who Solved the Greatest Scientific Problem of His Time*, Fourth Estate, London.

Spinoza, B. (2004), *Ethics*, Penguin Classics, London.

Squire, L. R. (1986), 'Mechanisms of memory', *Science*, 232, 1612–19.

Stephenson, L. M. (1967), 'A possible annual variation of the gravitational constant', *Proceedings of the Physical Society*, 90, 601–4.

Stevenson, I. (1997), *Where Reincarnation and Biology Intersect*, Praeger, Westport, CT.

Stier, K. (2010), 'Curbing drug-company abuses: are fines enough?', *Time*, 30 May http://www.time.com/time/business/article/0,8599,1990910,00.html

Strawson, G. (2006), 'Realistic monism: why physicalism entails panpsychism', *Journal of Consciousness Studies*, 13, 3–31.

Suzuki, D. T. (1998), *Studies in the Lakavatara Sutra*, Munshiram Manoharlal Publishers, New Delhi.

Tarnas, R. (1991), *The Passion of the Western Mind*, Harmony Books, New York.

Tegmark, M. (2007), 'The multiverse hierarchy', in Carr (ed.) (2007).

Temel, J. S., Greer, J. A., Muzikansky, A., Gallagher, E. R., Admane, A., Jackson, V. A., Dahlin, C. M., Blinderman, C. D., Jacobsen, J., Pirl, W. F., Billings, J. A., and Lynch, T. J. (2010), 'Early palliative care for patients with metastatic non-small-cell lung cancer', *New England Journal of Medicine*, 363, 733–42.

Thom, R. (1975), *Structural Stability and Morphogenesis*, Benjamin, Reading, MA.

Thom, R. (1983), *Mathematical Models of Morphogenesis*, Ellis Horwood, Chichester.

Thompson, E., Palacios, A., and Varela, F. J. (1992), 'Ways of coloring: Comparative color vision as a case study for cognitive science', *Behavioral and Brain Sciences*, 15, 1–26.

Thomson, W. (1852), 'On a universal tendency in nature to the dissipation of mechanical energy', *Proceedings of the Royal Society of Edinburgh*, 19 April.

Thurston, H. (1952), *The Physical Phenomena of Mysticism*, Burns Oates, London.

Time (1952), 'Medicine: entranced skin', *Time*, 1 September.

Trachtman, P. (2000), 'Redefining robots', *Smithsonian Magazine*, February, 97–112.

UK Government (2010), *Healthy Lives, Healthy People*, HM Stationery Office, London.

UK Office of Science and Technology (2000), *Science and the Public: A Review of Science Communication and Public Attitudes to Science in Britain*, UK Department of Trade and Industry, London.

Ullman, M., Krippner, S. and Vaughan, A. (1973), *Dream Telepathy Experiments in Nocturnal ESP*, Macmillan, New York.

Van der Post, L. (1962), *The Lost World of the Kalahari*, Penguin, London.

Velmans, M. (2000), *Understanding Consciousness*, Routledge, London.

Venter, C. (2007), *A Life Decoded*, Allen Lane, London.

Viveiros de Castro, E. B. (2004), 'Exchanging perspectives: the transformation of objects into subjects in Amerindian ontologies', *Common Knowledge*, 10, 463–84.

Waddington, C. H. (1957) *The Strategy of the Genes*, Allen and Unwin, London.

Wallace, B. A. (2009), *Mind in the Balance: Meditation in Science, Buddhism and Christianity*, Columbia University Press, New York.

Wallace, W. (1911), 'Descartes', *Encyclopaedia Britannica* (11th ed.), Cambridge University Press, Cambridge.

Wallace, A. R. (2000), *The Taboo of Subjectivity*, Oxford University Press, Oxford.

Watkins, A. J., Goldstein, D. A., Lee, L. C., Pepino, C. J., Tillett, S. L., Ross, F. E., Wilder, E. M., Zachary, V. A., and Wright, W. G. (2010), 'Lobster Attack Induces Sensitization in the Sea Hare, *Aplysia californica*', *Journal of Neuroscience*, 30, 11028–31.

Watson, J. D., and Crick, F. H. C. (1953), 'A structure for deoxyribose nucleic acid', *Nature*, 171, 737–8.

Watson, P. (1981), *Twins: An Investigation into the Strange Coincidences in the Lives of Separated Twins*, Hutchinson, London.

Watt, C., and Nagtegaal, M. (2004), 'Reporting of blind methods: an interdisciplinary survey', *Journal of the Society for Psychical Research*, 68, 105–14.

Webb, P. (1980), 'The measurement of energy exchange in man: an analysis', *American Journal of Clinical Nutrition*, 33, 1299–1310.

Webb P. (1991), 'The measurement of energy expenditure', *Journal of Nutrition*, 121, 1897–1901.

Weber, R. (1986), *Dialogues with Scientists and Sages: The Search for Unity*, Routledge & Kegan Paul, London.

Wegner, D. (2002), *The Illusion of Conscious Will*, MIT, Cambridge, MA.

Weil, A. (2004), *Health and Healing: The Philosophy of Integrative Medicine*, Houghton Mifflin, Boston, MA.

Weiss, P. (1939), *Principles of Development*, Holt, New York.

Westfall, R. S. (1980), *Never at Rest: A Biography of Isaac Newton*, Cambridge University Press, Cambridge.

Whitehead, A. N. (1925), *Science and the Modern World*, Macmillan, New York.

Whitehead, A. N. (1954), *Dialogues of Alfred North Whitehead*, Little, Brown, Boston.

Whitehead, A. N. (1978), *Process and Reality: An Essay in Cosmology*, Free Press, New York.

Whitfield, J. (2004), 'Telepathy charm seduces audience at paranormal debate', *Nature*, 427, 277.

Wicherts, J. M., Borsboom, D., Kats, J., and Molenaar, D. (2006), 'The poor availability of psychological research data for reanalysis', *American Psychologist*, 61, 726–8.

Wilber, K. (ed.) (1982), *The Holographic Paradigm and Other Paradoxes*, Shambala, Boulder.

Wilber, K., (ed.) (1984), *Quantum Questions*, Shambala, Boulder.

Will, C. (1971), 'Relativistic gravity in the solar system II. Anisotropy in the Newtonian gravitational constant', *Astrophysical Journal*, 169, 141–55.

Willis, A. (2009), 'Immortality only 20 years away says scientist', *Daily Telegraph*, 22 September.

Wilsdon, J., Wynne, B., and Stilgoe, J. (2005), *The Public Value of Science: Or How to Ensure That Science Really Matters*, Demos, London.

Winer, G. A., and Cottrell, J. E. (1996), 'Does anything leave the eye when we see?', *Current Directions in Psychological Science*, 5, 137–42.

Winer, G. A., Cottrell, J. E., Gregg, V. A., Fournier, J. S., and Bica, L. A. (2002), 'Fundamentally misunderstanding visual perception: Adults' beliefs in visual emissions', *American Psychologist*, 57, 417–24.

Winer, G. A., Cottrell, J. E., Karefilaki, K. D., and Gregg, V. A. (1996), 'Images, words and questions: Variables that influence beliefs about vision in children and adults', *Journal of Experimental Child Psychology*, 63, 499–525.

Wiseman, R. (2011), *Paranormality: Why We See What Isn't There*, Macmillan, London.

Wiseman, R., Smith, M., and Milton, J. (1998), 'Can animals detect when their owners are returning home? An experimental test of the "psychic pet" phenomenon', *British Journal of Psychology*, 89, 453–62.

Wiseman, R., Smith, M., and Milton, J. (2000), 'The "psychic pet" phenomenon: A reply to Rupert Sheldrake', *Journal of the Society for Psychical Research*, 64, 46–9.

Wiseman, R., and Watt, C. (1999), 'Rupert Sheldrake and the objectivity of science', *Skeptical Inquirer*, 23 (5), 61–2.

Woit, P. (2007), *Not Even Wrong: The Failure of String Theory and the Continuing Challenge to Unify the Laws of Physics*, Basic Books, New York.

Wolf, F. A. (1984), *Star Wave*, Macmillan, New York.

Wolpert, L., and Sheldrake, R. (2009), 'What can DNA tell us? Place your bets now', *New Scientist*, 8 July.

Wood, D. C. (1982), 'Membrane permeabilities determining resting, action and mechanoreceptor potentials in *Stentor coeruleus*', *Journal of Comparative Physiology*, 146, 537–50.

Wood, D. C. (1988), 'Habituation in *Stentor* produced by mechanoreceptor channel modification', *Journal of Neuroscience*, 8, 2254–8.

Woodard, G. D., and McCrone, W. C. (1975), 'Unusual crystallization behavior', *Journal of Applied Crystallography*, 8, 342.

World Health Organization (2003), *Acupuncture: Review and Analysis of Reports on Controlled Clinical Trials*, World Heath Organization, Geneva.

Wright, L. (1997), *Twins: Genes, Environment and the Mystery of Identity*, Weidenfeld and Nicolson, London.

Wroe, A. (2007), *Being Shelley: The Poet's Search for Himself*, Vintage Books, London.

Yates, F. A. (1969), *The Art of Memory*, Penguin, Harmondsworth.

Young, E. (2008), 'Rewriting Darwin: the new non-genetic inheritance', *New Scientist*, 9 July.

Zajonc, A. (1993), *Catching the Light: The Entwined History of Light and Mind*, Bantam Books, New York.

Zhang, B., Wright, A. A., Huskamp, H. A., Nilsson, M. E., Maciejewski, M. L., Earle, C. E., Block, S. D., Maciejewski, P. K., and Prigerson, H. G. (2009), 'Healthcare costs in the last week of life', *Annals of Internal Medicine*, 169, 480–88.

Ziman, J. (2003), 'Emerging out of nature into history: the plurality of the sciences', *Philosophical Transactions of the Royal Society A*, 361, 1617–33.

Index